Praise for *An End to Upside Down Living*

"Mark Gober's first book, *An End to Upside Down Thinking*, provided clear presentation of the emerging scientific revolution surrounding the mind-brain connection and its world-changing implications for more fully understanding reality. In this absolutely essential sequel, he offers extraordinarily rich insights and practical applications stemming from these advanced ideas supported by deep conceptual background. This remarkable series energizes a profound revolution in how we can *live* more authentic and purposeful lives, enriching our individual and collective experience beyond imagination. Highly recommended!"

—**Eben Alexander, MD**, neurosurgeon and author of *Living in a Mindful Universe, Proof of Heaven*, and *The Map of Heaven*

"In *An End to Upside Down Living*, a follow-up to his important *An End to Upside Down Thinking*, author Mark Gober turns his attention from thinking to actual living. How does one put these vital ideas into practice? What difference do they make in one's life? How did he, himself, accomplish this—and why are these steps necessary for our survival as a species? This is no clever iconoclastic outburst, but a deeply sane, honest, and vital handbook for survival. We ignore Gober's voice at our own risk."

—**Larry Dossey, MD**, Author: *One Mind: Why Our Individual Mind Is Part of a Greater Consciousness and Why It Matters*

"[Gober] takes the upside down world of business and of modern life in general, and puts them back on sound and sane feet. Worth reading—and taking to heart!"

—**Dr. Ervin Laszlo**, Nobel Peace Prize nominee and systems theorist

"Knowledge based on faulty or incomplete paradigms is necessarily flawed (geocentric astronomy, medicine before germ theory, physics before Newton, Einstein, and Schrödinger). If they consider consciousness at all, modern sciences assume that it is an epiphenomenon of brain functioning. If, as Mark Gober convincingly argues, consciousness is the foundation of the universe, then it is the foundation of all fields of knowledge, as well as their applications. Yet, aside from a few outliers, no modern discipline recognizes this. Consequently, our foundationless quest for understanding has brought us great benefits but also great harm and possible extinction. If consciousness is fundamental, then understanding it and harnessing its full potential will be humanity's greatest breakthrough. This, therefore, may be one of the most important books you'll ever read."

—**Rick Archer**, host of *Buddha at the Gas Pump*

"This book is an easy and enjoyable read, which is saying something given that it addresses most of the major mysteries of human existence and the nature of reality! It's a book that is thought-provoking, inspirational, and down-to-earth all at once. Whether you digest it slowly, reflecting on the premises made in each chapter one at a time, or devour it in one sitting, it's sure to shift your perspective and help to calibrate how you choose to view the world, and in turn, how you choose to live."
—**Cassandra Vieten, PhD,** Senior Fellow and Past President, Institute of Noetic Sciences; and Scholar-in-Residence, Arthur C. Clarke Center for Human Imagination, University of California, San Diego

"For spiritual travelers, Mark Gober's newest book is like getting a ticket for an evolutionary soul journey to the infinite field of unified awareness! Readers become true seekers, as they soon discover what they thought they know, may not be their own truth anymore. They learn about 'unknowns' and then realize that there are even more 'unknown unknowns' on this trip through life. I found this an enlightening and intelligent presentation that fully supports my own 'One Consciousness' experiences."
—**William "Rev. Bill" McDonald Jr.,** Author: *Warrior: A Spiritual Odyssey* and *Alchemy of a Warrior's Heart*; award winning poet and a three-time near-death experiencer

An End to Upside Down Living

Also by Mark Gober

AN END TO UPSIDE DOWN THINKING

Dispelling the Myth That the Brain Produces Consciousness,
and the Implications for Everyday Life

An End to Upside Down Living

Reorienting Our Consciousness to Live Better and Save the Human Species

Reorienting Our Consciousness to Live
Better and Save the Human Species

Mark Gober

Waterside Productions
Cardiff-by-the-Sea, California

Waterside Productions
2055 Oxford Avenue
Cardiff-by-the-Sea, CA 92007
www.waterside.com

To those in search of Truth. And those aspiring to live in accordance with it.

"We are all caught in an inescapable network of mutuality, tied in a single garment of destiny."

—Rev. Dr. Martin Luther King Jr.

CONTENTS

SETTING OUR COMPASS

I have a simple but huge question for you:

What is the overall intention of your life?

Think about that for a few moments.

While you're thinking about the answer, I'll give more context. Another way of phrasing it is: What is the orientation that fundamentally drives all of your life's values, priorities, and decisions; and ultimately, what you do in the world?

How we *intend* to live is the essential guiding force behind the life we *do* live. The direction in which we orient our compass determines where we might end up.

Imagine two boats. They start off at the same dock, but they orient their respective compasses a mere one degree differently. In the first few minutes of sailing, the boats are pretty close to each other. After a few hours, they're farther apart. The distance grows significantly as time goes on, until the two boats are sailing in completely different oceans. The initial, seemingly small, difference ultimately results in vastly different trajectories.

Now let's apply that metaphor to life. The compass driving our own lives needs to be finely tuned in order to have a chance at achieving our desired outcomes. Even tiny tweaks in our orientation can push us toward a very different life. And moreover, big changes in our orientation can result in massive changes in our lives.

The purpose of this book is to explore, with precision, where we should set our life's compass. What do I mean by "should"? I mean that I'm looking for an orientation that most closely aligns with the true nature of reality. In other words, this exercise requires having a perspective on what life is, who and what we are, what matters, and why. Our orientation is set on the basis of those foundational beliefs.

My research has led me to reason that our collective orientation, as a civilization, is completely off the mark due to a deep scientific misunderstanding of reality. We aren't just one degree off, either. We aren't 45 or 90 degrees off. We're closer to 180 degrees off.

It all boils down to the following idea, which we'll explore soon in detail: **The overarching belief in separation—rather than interconnectivity—is leading to misguided priorities and decisions across the globe.** Virtually all of the world's countless problems can be linked back to this misdirected compass. The increasingly disturbing external world is a mirror of our combined internal compasses. And so, our collective boat has veered into lethal waters. For this reason, I've written each word of this book with a genuine sense of urgency.

Now you might understand more viscerally why I devoted an entire book, *An End to Upside Down Thinking* (2018); and a podcast series, *Where Is My Mind?* (2019), to correcting the way we scientifically view reality. In my own life, a fundamental shift in *thinking* about reality has led to drastic, positive changes in practical *living*.

In this book I aspire to provide you with a road map for making similar changes in your life—to arm you with an approach for setting your life's overall intention *in a way that aligns with reality*.

And as each of us does so individually, we open the possibility of collectively redirecting our civilization's boat and uplifting the planet before it's too late. In order to shift the external, we need to shift the internal. I view this exercise as the only hope for the survival of the human species as the clock ticks ominously.

Mark Gober
February 2020
San Francisco, California

CHAPTER 1

INTRODUCTION

My Upside Down Life

You might be wondering how I would answer the question I posed in the preface: *What is the overall intention of your life?*

If you had asked me that question several years ago, I would have had a simple answer: *The question itself is irrelevant because life has no intrinsic meaning.*

By the end of this book, we'll get to my current answer. I'm intentionally not giving it now because without proper context it won't fully make sense. In order to develop a precise orientation for our lives, we need to build up to it by clarifying what reality is and what type of living approach logically follows. That's what this book is about.

We'll begin with my own path and why I felt life had no meaning (chapter 1). We'll then transition to examining the evidence that forced me, albeit reluctantly, to accept a completely different worldview (chapter 2). That worldview leads to an important set of inferences we can make about the nature of reality and our place in it (chapter 3), which then informs ways in which we might approach life accordingly (chapter 4). This overall attitude places us on a well-traveled personal growth path (chapter 5), requiring discernment and an awareness of common pitfalls (chapter 6). Finally, with that backdrop, I lay out my life's new orientation and why pointing our compass in that general direction is needed to improve our individual lives and save the species (chapter 7).

My upbringing

I was never a religious person. I grew up in a conservative Jewish household where we celebrated the major holidays as a family, but not much else. My grandfather, who lived nearby and was a big part of my upbringing, was a Holocaust survivor. As a result of his experiences escaping Nazi Germany, he felt a sense of responsibility to preserve Jewish traditions. Out of immense respect for him, I always went through the motions—even getting a bar mitzvah (a Jewish coming-of-age ceremony at age 13)—but otherwise I had no real interest in any of it. My knowledge of religion and spirituality was always very "surface level." I knew the basics but hadn't studied any of the traditions in depth, nor did I have any interest in doing so.

Throughout my education—first at one of the nation's top prep schools located in Baltimore, Maryland, and then at Princeton University—I was taught the skill of critical thinking. The directive to question *everything* was hammered into my psyche in every academic corner. And yet when it came to religion (Judaism or otherwise), I saw that people were adhering to ancient scriptures, written by who knows who, who knows when. And who knows how much of those writings had been changed or manipulated over the years to support individual agendas? So I was naturally skeptical of religion. I would think to myself, *Why do we critically evaluate ideas in all domains, but when it comes to religion, we're just supposed to blindly accept without scientific evidence?* On top of that, I had no direct experience of the stories religions described. God never spoke to me through a burning bush. I never saw divine beings or witnessed miracles. How was I to relate to any of that? And where did God even come from?

To me, religions were convenient belief systems stemming from our human desire to find meaning, even if no meaning actually existed. Notions of an afterlife or reincarnation seemed like methods for comforting ourselves in the face of our inevitable mortality.

But the worst part of religion for me was this: If each religion was purporting something different, then, I reasoned, they couldn't *all* be right. Therefore, I felt that in the most optimistic scenario, all religions but one (*if* any one was in fact correct) wagered everything on an incorrect worldview. And these incorrect worldviews have been at the heart of countless murders and conflicts throughout human history—in the name of serving religious beliefs with zero proof.

Religion claimed to provide moral and ethical codes, but even that didn't seem right to me. We didn't need religion to tell us how to act; our own evolutionary impulses would naturally do the trick. For example, former Oxford biologist Richard Dawkins's "selfish gene" theory explains how morality is a natural by-product of evolution: when you act well toward others, they in turn are more likely to act well toward you, thereby giving you a better chance to survive in order to reproduce. Our genes' own "selfish" desire to persist selects for unselfish behavior toward others.

It's probably not surprising to hear that one of my favorite books used to be Dawkins's *The God Delusion* (2006). While I still remain critical of many aspects of religion, I now realize that I was throwing the proverbial "baby out with the bathwater" when it came to notions of "God," the "supernatural," and the broader category of "spirituality."

A meaningless universe

In hindsight, I recognize that the most fundamental reason for dismissing these notions was a belief in the metaphysical philosophy of **physicalism**. It is the current meta-paradigm—the foundational paradigm underlying all other paradigms for much of science and everyday living. (Note: In my previous book, I referred to this philosophy as *scientific materialism*. I'm not using that term in this book because sometimes the term is misunderstood. Scientific materialism does not refer to the overvaluing of material possessions. To avoid confusion, I'm using *physicalism* instead, which is essentially the same as scientific materialism.)

Physicalism is so deeply ingrained in our education system and Western culture that it often goes unacknowledged as the foundational framework for thinking about life. I don't recall hearing the term used even once when I was a student at Princeton. And yet it served as the underpinning of basically everything I was taught, and it shaped how I viewed my identity.

Physicalism suggests that the universe is fundamentally made of physical stuff called "matter." As the theory goes: Roughly 13.8 billion years ago, somehow, a "Big Bang" started the universe and filled it with units of matter ("atoms"). Biological organisms, such as human beings, evolved over many, many years, out of the random combinations of atoms (via chemistry). The theory is: Given enough time and enough random chemical reactions in a "primordial soup" of matter, a self-replicating molecule (for example, DNA) could form by accident. Its emergence is highly improbable, but when dealing with hundreds of millions of years, it could happen. Dawkins offers a football-pool analogy to explain this: If one were to partake in a football pool "every week for a hundred million years you would very likely win several jackpots.... [And] it only had to arise once."[1] Under physicalism, we did hit a jackpot, and life emerged.

The random combinations of matter have no real meaning behind them, however. So physicalists, if they're being totally honest with themselves, are forced into a bleak outlook on life. As Nobel Prize–winning physicist Steven Weinberg said: "The more the universe seems comprehensible, the more it also seems pointless."[2] I respect Weinberg's honesty here. The more educated I became, the more I bought into this stance.

On the other hand, I was disturbed by physicalist/atheistic scientists who tried to *manufacture* meaning. For example, Caltech physicist Dr. Sean Carroll states: "Our emergence [through evolution] has brought meaning and mattering into the world."[3] To me, this is similar to the religionist who is fishing for meaning through a leap-of-faith belief in "God" as a subliminal way to comfort him or herself. Carroll's statement is a rationalization. It's *creating artificial meaning* in a universe that has no *intrinsic meaning.*

Author Alex Tsakiris, host of the *Skeptiko* podcast, stated it well: "If the universe is meaningless, and you are embedded in that meaningless universe, then there can be no meaning to your life. It can't be any other way. If there's real meaning in/to your life, then there's at least that much meaning in the universe. If the universe is meaningless, if there isn't even a tiny little smidgen of it anywhere in the universe, then you have zero chance of ever finding it in your life."[4]

This is the brutally honest reality a physicalist has to come to grips with. I accepted this notion because I didn't want to rationalize. I wanted to accept reality the way it was, whether I liked it or not.

Furthermore, I often encountered commentary from the modern-day "Four Horsemen" of atheism—Christopher Hitchens, Daniel Dennett, Richard Dawkins, and Sam Harris. They were at times militant about their atheistic views. Although I didn't disagree with their philosophical positions, I couldn't understand why they were so vocal and fervent. It seemed hypocritical. I would think to myself, *Your worldview implies a meaningless universe, so why are you spending so much time and energy talking about it? If life has no meaning, it's even meaningless to argue about it. Why do you guys care so much about being right?* Philosopher Alan Watts summed it up concisely: "If the universe is meaningless, so is the statement that it is so."[5]

Consciousness

Physicalism also contemplates the origin of "consciousness." Consciousness is the part of us that experiences life; it is the formless awareness that looks through our eyes and hears through our ears. Under physicalism, consciousness is part of the accidental and random evolutionary process. It emerges out of our brain (sometimes known as an *epiphenomenon*). So it is only through the evolution of a brain that we have the capacity for consciousness[6] [image follows, adapted from Dr. Dean Radin]. If life hadn't emerged out of the primordial soup of matter, there wouldn't be any consciousness in the universe.

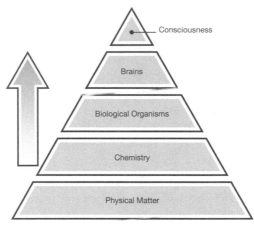

Physicalism: The interactions of units of matter (via chemistry) create biological organisms like human beings, which develop brains, out of which consciousness arises.

The belief that the brain creates consciousness is completely unproven. More than that, scientific evidence suggests that this belief is incorrect (as I showed in my previous book and podcast). Instead, the evidence suggests that consciousness exists independently of the body and is the basis of reality. So, matter is actually derived from consciousness, which is the opposite of what physicalism is saying (more on this in chapter 2).

For now, let's dig into another major implication of the physicalist belief system. If, as physicalism claims, the brain is responsible for our sense of experiencing life, then what happens when the brain stops functioning (that is, at death)? Our experience of life is over. Our consciousness is gone because the machinery that creates consciousness—the brain/body system—isn't functioning. It's like shutting a computer off. Lights out. When you die, it's over. No memories. No emotions. No thoughts. Some people push back on this, claiming, "Well, we might have offspring, family, and friends who are alive even after we die." But the harsh reality, under physicalism, is that anyone we leave behind will eventually die too. The only question is when and how the inevitable will happen for every living creature. As my high school biology teacher put it, "Life is a terminal illness."

I'm intentionally belaboring this point because how we view death is *essential* for how we think about meaning in life. I feel this idea is drastically overlooked in our education system and in our everyday thinking. It's the "elephant in the physicalist's room": whatever artificial meaning someone might have created while living—which didn't even matter to begin with—is wiped out when the person dies.

So as a physicalist, what is the prescription for how you should live your life? How can you even contemplate an overall life intention or orientation when mired in meaninglessness? There is nothing with true meaning to direct the compass. This was exactly my conundrum.

My journey under the premise of meaninglessness

These were not topics I spoke about much. First of all, in my social circles this level of conversation didn't come up often. Second, it's morbid. But my nihilistic worldview always lingered in the back of my mind.

Yet, for some reason, I was motivated in life. For example, in high school I earned the highest grade-point average my junior year, was elected class president, and was a nationally ranked tennis player—all while managing an active social life. I barely slept because I was so busy. I didn't have a sense of why "success" mattered, but I felt a self-imposed compulsion to thrive. It wasn't optional, either. I had to succeed at the highest level in *everything*.

My path as an aspiring achiever didn't end in high school. I was accepted into Princeton University against statistical odds. Not only was it the #1-ranked university, but I was given a spot on the school's Division I tennis team coached by a former world #1-ranked professional doubles player. My freshman year (2004) was the first in which Princeton instituted a "grade deflation" policy, meaning the administration capped the number of "A" grades it awarded. So in leaving the pressure cooker of my elite high school, I took things to the next level in college and continued my workaholic ways.

I managed to do well, graduating *magna cum laude* with a degree in psychology (emphasizing behavioral economics) and wrote an award-winning senior thesis on Daniel Kahneman's Nobel Prize–winning "Prospect Theory." Also, I was elected a captain of the tennis team my senior year. Yet, while managing a brutal academic schedule on top of a semiprofessional athletic commitment, I had no idea why any of it mattered. I just tacitly assumed that if I kept succeeding, eventually I'd reach a mythical point of lasting happiness and satisfaction.

That's the way my peers seemed to be treating life. It was almost as if people were living by the following philosophy: **Get a good education; make a lot of money; get married; have kids; take care of your kids; die.** The implicit mantra was that "he or she who dies with the most toys wins." Material success was the force driving one's motivations, and the underlying meaning of life was an afterthought that one could look into for fun in retirement. I didn't necessarily buy into this attitude, but it was the mentality I saw just about everywhere I looked.

After graduating from Princeton, I didn't slow down. I chose one of the most intense professional environments possible: Wall Street. I was hired to work at a large global investment bank in New York, which I joined in the summer of 2008. I was there during the heart of the financial crisis, tasked with advising struggling financial institutions. I survived rounds of layoffs and worked nonstop. I went for consecutive months without a day off, including weekends. When I first started at the firm, my standard for a "good day" was if I left the office before 2 A.M.

To say I disliked that job would be an understatement. One evening, a colleague and close friend jokingly drew a rendition of Dante's *Inferno* (the layers of hell) on the conference-room whiteboard to represent where we were. But my job was considered such a coveted position that employees simply didn't jump ship without good reason. People seemed to have "golden handcuffs" and knew there were countless others who wished they had an opportunity to be in such an elite role.

Eventually my job in investment banking became a health hazard. My body was starting to shut down. I recall one evening in the fall of 2009 when I was working on a financial model at my desk. One of my supervisors was up late with me, breathing down my neck, making sure we were fine-tuning the model to perfection. Suddenly, my heart began beating irregularly, and for a moment I passed out. The next day I saw my doctor, and he told me, "Your heart is fine, but you need to change your lifestyle and take it easy."

I knew it was time to leave my job. Not only was I suffering physically and emotionally, but I wasn't enjoying the work itself. I didn't see value in it beyond hopefully jettisoning my career into more rapid success. One of my New York roommates, a friend from Princeton, lightheartedly told me, "You're a lost soul."

In 2010, I left Wall Street and moved to Boston for my new job, joining what is now called Sherpa Technology Group. The firm advises both large and small corporations on strategy and mergers/acquisitions for technology and intellectual property (that is, patent) matters. It was founded by world-leading experts on the topic. I was excited about the opportunity to work with the best of the best on cutting-edge technology projects and develop specialized knowledge about the intersection of business, technology, and law.

After working hard for 18 months in the firm's Boston office, I was offered a role in the Silicon Valley location. I excitedly moved to San Francisco and immediately began working on high-profile projects. Within a year of making the move, I had the rare opportunity to give presentations to boards of directors of two publicly traded companies that were in distress and that needed my (and my colleagues') advice for their survival. It was quite an experience to be so young and presenting to seasoned professionals who were decades older than I was.

This all came at a cost, however. My career continued to dominate my life. Even when I wasn't working, I was spending my evenings and weekends teaching myself patent law and getting into the weeds of technology. It was all very interesting, so I was okay with it. I still had no sense of why any of it fundamentally mattered, though.

As I progressed at the firm, reporters (for example, at *Bloomberg Businessweek*) began asking for my thoughts on business matters and often quoted me. I also wrote internationally published articles. One of our client CEOs even mentioned mc during a public presentation to investors, referring to me as "one of the sharpest young individuals I've worked with." I was having an amazing professional learning experience, with exposure that I couldn't have anticipated, incredible mentorship from my managers, and promotions along the way. On paper, things looked pretty good.

Life crisis

In 2014/2015 a life crisis hit. The patent landscape was beginning to change based on significant legal and regulatory shifts. The outcome was that patents became difficult to enforce against infringers. Companies could suddenly infringe upon competitors' patents with less risk, and that naturally reduced the overall value of patents. This led to the "other great recession" that most people haven't heard of: the great patent recession. As a result, and for reasons completely beyond my control, a number of business deals I had poured extraordinary effort into failed. (If only the details weren't confidential, these stories alone would make for a fascinating book! They involved multibillion-dollar courtroom dramas, stock-price manipulations, defamation, innovation theft, and more.)

On top of these professional blows, after years of being laser focused on my career, I started dabbling with dating. None of these relationships were very serious, but the ones that ended were disappointing even though I knew the women weren't ultimately right for me.

At this point in my life, I really did feel like a lost soul. I wasn't sure where my career was heading, and I was uncertain that I would ever find a romantic relationship. Basically, I had no direction in my life, and lying dormant in the background was the belief that life had no meaning anyway. I spent the 2015 winter holidays with my family and wasn't able to get out of bed for days. I don't know if anyone knew exactly what was wrong. In hindsight, I don't think I fully understood either. I was in a hopeless state, unhappy with my life, about to turn 30, and had no idea what could make things better. I felt like the living dead: a true zombie.

Because I knew of no other option, I powered through this intensely awful period and went through the motions at work. Thinking back on that time, it was as if my lens on life was clouded with a haze of despair that simply wouldn't go away.

A shift

In August 2016, the first signs of a shift started taking place. Earlier that year I had begun listening to podcasts, primarily on the topics of business and health. One of the shows I found interesting was called *Extreme Health Radio*. I randomly ended up listening to an interview recorded months before with a woman named Laura Powers. She sounded like a "normal" person, holding a master's degree in political science, and she had worked in higher education.

She spoke about having alleged "psychic" abilities and advised clients by using these skills. That show alone didn't shift any of my perspectives, but I was interested enough to subscribe to Powers's own podcast called *Healing Powers*. She interviewed people who had experiences similar to hers. The interviews were relatively short and easy to listen to during my long daily commute in Bay Area traffic. I simply let the podcast play in the car, one interview after the next. I was mostly just entertained by hearing about new topics.

After a few weeks of listening, entertainment transformed into intrigue. It dawned on me that many of the people Powers interviewed were *independently* describing a view of reality that was completely counter to everything I'd been taught. In other words, they were challenging physicalism. I kept listening. I had a hard time reasoning that all of these people were delusional and/ or blatant liars. I needed more than anecdotal evidence, so I investigated elsewhere. I read books, listened to other podcasts, read scientific papers, and even had my own sessions with psychics. Eventually, I spoke with scientists directly.

It's difficult to put into words how disorienting this period was for me. Everything I was learning was pointing toward a totally new way of looking at life—one in which the "supernatural" was simply part of human reality. The more I researched, the more the evidence converged in that direction. The even bigger implication of my research was that life *did* have some intrinsic meaning. I was forced to rethink what "reality" actually was and what it meant to be human. I had no choice but to reconsider all my assumptions about life. In spite of having had what I thought was a strong education, at this point I realized that I had been going through life in a state of unimaginable ignorance.

Over the Thanksgiving holiday that year, I decided to stay in the Bay Area by myself rather than be with my family. I spent several days in nature, convening with the ancient trees in Muir Woods. This quiet time gave me an opportunity to think about how to live life with a new perspective and determine whether I had simply lost my mind.

It was surreal to go from a place of abject hopelessness to being slapped in the face with a completely new worldview. To be clear, I was *not* consciously looking for this. I wasn't saying: *Life isn't going well; I'll find a new worldview so I can manufacture meaning and comfort myself.* I just stumbled on the path and caught wind of how big it was. It was like starting life all over again.

My research continued for months. By "research," I mean it's *all* I did when I wasn't working at my day job. This was all I wanted to think about. It was a back-and-forth process: I would learn an amazing fact and realize the implications, and then mundane reality would set in again and I would forget. Then I would learn more and become excited. The process continued like that for a while. My perspective began to shift, and life slowly got better.

In June of 2017 before driving to work one day, I remember having an insight: *I should summarize my findings in a book.* At first, I resisted. Making that happen would take so much effort. And even more than that, how could a business guy like me write a book about paranormal phenomena that challenges all of modern science? Would anyone listen? And would people think I'm crazy?

Those concerns didn't last long, however. I didn't care what people thought because I would be conveying the information through data and credible science. If they wanted to disagree, let them disagree with all the scientists. I was just the messenger. Most important, I knew the information could—and would—transform other people's lives in the same way it had transformed mine. I felt I had no choice but to summarize my findings and make the information accessible to others. After a nice nudge from two buddies over dinner in June 2017, I realized that the book simply had to be written.

Once I made the decision to write a book, I decided I wanted to finish it immediately. I had researched extensively for almost a year, so I had a general sense of its content. I then harnessed my inner investment banker and worked on the book nonstop over the July 4 holiday weekend. Over the next few weeks, in between my business obligations, I finished the complete manuscript for *An End to Upside Down Thinking*, which was eventually published in 2018. Since then, I've launched the podcast *Where Is My Mind?* (2019), which features my interviews with dozens of world-leading experts. Additionally, I've become more personally involved in the scientific exploration of these topics and have joined the board of directors of the Institute of Noetic Sciences (IONS). IONS is a consciousness-focused research center with a campus in Petaluma, California, founded in 1973 by former Apollo 14 astronaut Dr. Edgar Mitchell. I also joined the board of a new spiritual education platform and retreat center that will be based outside of Asheville, North Carolina (the School of Wholeness and Enlightenment, or SoWE). On top of all that, I still research and ponder these topics nonstop.

Most important, the "haze" that was temporarily clouding my life has been lifted. I now live with a much clearer sense of purpose and meaning. The rest of this book aims to reveal the thinking that led me to this place, which has translated into a revitalized manner of living. If I had known about the information presented in these chapters years ago, I could have avoided so much suffering. I sincerely hope it will be of assistance to you.

CHAPTER 2

UPROOTING ASSUMPTIONS

A New Meta-Paradigm

Looking back, I now realize that my personal downturn was inevitable because of my worldview. My underlying physicalist belief system left no room for finding meaning in life. No matter how much I achieved, I never experienced lasting peace or happiness because none of my achievements had fundamental meaning. I was endlessly seeking the "next best thing," effectively running on a treadmill. Five minutes after one achievement, I was worried about the next one I needed to attain. And it seemed like everyone around me was on this very same path, hypnotized by conditioning and social norms that they rarely stopped to question.

My life was an example of how covertly devastating the physicalist worldview could be. Serendipitously, I found my way out. Others aren't so lucky. Writing this book felt like a responsibility to my fellow humans to show that there *is* an alternative way of living that actually aligns more closely with reality. I will repeat that for readers who are struggling with life: **there is an alternative!**

In my previous book and podcast, I aggregated an abundance of scientific evidence suggesting that physicalism is wrong. Instead, the meta-paradigm of reality I endorse is one that happens to align with what many call "spiritual"—not because I'm looking for a way to comfort myself, but because *that's simply where the science points*. The term *spiritual* refers to the notion that there's a part of us that transcends our individual identity as physical beings.

More specifically, my research has focused on the assumption that consciousness is produced by the brain. Here, I will summarize the key findings that will serve as a foundation for the rest of this book. However, this intentionally brief overview is, in my opinion, no substitute for the full onslaught of scientific evidence provided by my previous book, podcast, and the associated source material. In addition to my book and podcast, a few excellent books on this topic include: *One Mind* (2013) by Larry Dossey, MD; *Transcendent Mind* (2017) by Imants Barušs, PhD, and Julia Mossbridge, PhD; *Entangled Minds* (2006) by Dean Radin, PhD; *Irreducible Mind* (2009) and *Beyond Physicalism* (2015) by Ed Kelly, PhD et al.; *Mind Beyond Brain* (2018) by David Presti, PhD et al.; *The ESP Enigma* (2008) by Diane Powell, MD; *Consciousness Beyond Life* (2007) by Pim van Lommel, MD; *Living in a Mindful Universe* (2017) by Eben Alexander, MD and Karen Newell; *The Reality of ESP* (2012) by Russell Targ; *Connected* (2019) by Roger Nelson, PhD; *The Sense of Being Stared At* (2013) by Rupert Sheldrake, PhD; *The Cosmic Hologram* (2017) by Jude Currivan, PhD; and *Spiritual Science* (2018) by Steve Taylor, PhD.

The fatal assumption

Physicalism *assumes* that the brain creates consciousness because brain states correlate with conscious states. What does that mean? It means that when you change the brain, conscious experience changes. For example, if you damage someone's brain in the area responsible for vision, the person might have difficultly seeing. If someone has a neurodegenerative disease, like Alzheimer's disease, he or she might experience memory loss. The field of neuroscience focuses on identifying these connections between what happens to the brain and what happens to consciousness, known as "neural correlates of consciousness."

Unfortunately, however, many scientists are making an unwarranted assumption: they assume that because the brain is *related* to consciousness, it follows that the brain *creates* consciousness. This runs into the potential error often stated in statistics as "correlation does not imply causation." If two concepts are related, there is no guarantee that the relationship is *causal*.

An example I often reference comes from philosopher Dr. Bernardo Kastrup, who says that if you have a large fire, fire fighters appear to put it out. If you have a larger fire, more fire fighters show up. There's a strong correlation between the size of the fire and the number of fire fighters who appear at the scene. But does that mean that the fire fighters *caused* the fire? Of course not.[1] This is the logical error scientists are making with consciousness, and it's so incredibly costly.

Ignoring logic for a minute, there is also an open scientific secret: **No one has any idea how consciousness could come out of a brain.** I dare you to ask your local neuroscientist how the brain creates consciousness. See what he or she says. Neuroscience PhD Sam Harris summarizes it well: "There is nothing about a brain, studied at any scale, that even *suggests* that it might harbor consciousness"[2] [emphasis in original].

The issue is this: our consciousness is formless and subjective, whereas our brain is physical and tangible. You can touch your brain, but you can't touch your consciousness. How can something that you *can't* touch—like consciousness—emerge from something you *can* touch—your brain? No one knows! This is the infamous "hard problem" of consciousness, and *Science* magazine listed this question second on its list of the top twenty-five questions remaining in science.[3]

The brain as a "blindfold" of consciousness

The brain is, of course, related to the way we experience the world; I'm not disputing that. I'm simply arguing that we need to recontextualize the brain's role. Maybe the brain doesn't play the role of creating consciousness, as so many people have assumed.

I argue that the brain acts more like a filtering mechanism that restricts, limits, and processes a consciousness that exists independently of the body. Consciousness doesn't need a brain to be conscious. Rather, the brain gets in the way like a blindfold; it is the lens that governs the type of consciousness we experience. So when we get the brain out of the way, we can experience an enriched consciousness—one that has always been always there, but which our brain blocks us from experiencing fully.

There are a number of phenomena that align with this idea. Many of them are also discussed in Kastrup's excellent *Scientific American* article "Transcending the Brain" (2017).[4] These phenomena show the general pattern of "less brain, more consciousness," which is what you'd expect if the brain acted like a filter or blindfold.[5] Examples include:

○ *Savants:* Individuals with severe brain impairments sometimes exhibit extraordinary mental abilities (for example, math, memory, music, and so on). As an example, Kim Peek—the man the movie *Rain Man* was based on—was able to memorize the index of a set of encyclopedias at age six in spite of having major brain deficiencies.[6]

○ *Near-death experiences:* People with severe bodily trauma, such as cardiac arrest, sometimes report accurate memories that are *clearer than usual or more logical than usual,* in spite of having little (sometimes zero) brain functioning. As University of Virginia professor Dr. Bruce Greyson aptly put it: "We're left with this paradox that at a time when the brain isn't functioning, the mind is functioning better than ever."[7]

○ *Psychedelics:* Recent studies suggest that psychedelic experiences result in an enriched sense of reality accompanied by *reductions* in brain activity. Thus, a psychedelic could be viewed as a mechanism for "unlocking the brain's filter" to a reality that was always there, but which our brain normally blocks.[8]

○ *Terminal lucidity:* Patients with severe neurological disorders (for example, Alzheimer's disease) sometimes snap back into mental clarity shortly before dying. These are cases of a clear consciousness with a highly damaged brain.[9]

○ *Hydrocephalus:* Some people with almost no brain tissue (their brains are full of cerebrospinal fluid) are fully functional and intelligent. For example, a patient studied by Dr. John Lorber at Sheffield University lived a normal life and graduated with honors in mathematics (IQ of 126). The patient needed a CAT scan for a minor ailment, and to the doctor's great surprise, the man had virtually no brain.[10]

○ *Shufflebrain:* Indiana University anatomy professor Paul Pietsch mashed salamander brains, and yet they still exhibited feeding behavior. As a physicalist, Pietsch's own studies shocked him: "My very own research... forced me to junk the axioms of my youth and begin my intellectual life all over again."[11]

○ *G-force-induced loss of consciousness:* Blood is forced out of the brain, and yet people report "memorable dreams."[12]

○ *Holotropic breathwork:* A breathing technique using hyperventilation can cause expansions of awareness.[13]

○ *Choking game (dangerous!):* Through partial strangulation and fainting, people can induce feelings of transcendence.[14]

○ *Psychography:* Neuroimaging shows reductions of activity in key brain areas when an alleged psychic produces writing material (from a "transcendent" source). The writing is done in a trancelike state and is more complex than material produced without a trance.[15]

Whirlpools in a stream of consciousness

If we accept that the brain does not produce consciousness

and instead limits it, then we have to wonder where consciousness comes from. The view I'm most drawn to is how Nobel Prize–winning physicist Max Planck stated it in 1931: "I regard consciousness as fundamental. I regard matter as derivative from consciousness. We cannot get behind consciousness."[16]

Furthermore, consciousness exists beyond space and time as the basis of all reality. Nothing "caused" it to exist; it simply *is* without something causing it to be (as counterintuitive as that sounds). Notions of "linear time" and "causality" are merely illusions and human constructs. So the notion that something would have needed to cause consciousness is an artifact of the mind's limited capacity to understand reality. I'll discuss this in greater depth later on.

In essence, the model argues that we are all fundamentally interconnected as part of the same underlying consciousness. Nobel Prize–winning physicist Erwin Schrödinger summarized the idea well: **"In truth, there is only one mind."**[17] Below is a rough illustration of the idea.

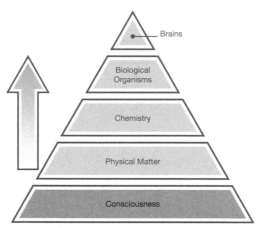

The "One Mind" view of reality
Note: This image is merely a visual approximation. If consciousness is truly the basis of reality, then everything else in the triangle is simply a modulation of consciousness itself.

Kastrup offered a metaphor to explain the basic idea further.[18] He said that we can imagine that all reality is one universal consciousness represented by a stream of water. Within the stream, whirlpools form. Each whirlpool has a sense of individuality, but they're interconnected as part of the same stream. Using another one of Kastrup's analogies, it's as if the stream of consciousness— **we'll call it the "One Mind"**—has dissociative identity disorder. Each of us is an individual "alter" of the whole.[19]

So, the reason the hard problem of consciousness hasn't been solved is that it's posing the wrong question. It's asking how a brain can make consciousness... when the brain doesn't make consciousness in the first place! No wonder *Science* magazine's #2 question has gone unanswered. And incidentally, this consciousness-centric view of reality answers *Science* magazine's #1 question in all of science: "What is the universe made of?" The answer, according to this framework: *consciousness*.

Back to whirlpools. Imagine that I'm one whirlpool and you're another. If some of the water escapes my whirlpool and enters yours, it's akin to saying that some of my consciousness enters yours. In other words, it's a telepathic or psychic occurrence. **So this model predicts that psychic phenomena would be real.** Furthermore, if a whirlpool delocalizes, the water flows back into the broader stream. But it doesn't leave the stream. The water simply changes form. Analogously, when the physical body dies, consciousness doesn't die but instead transitions into a new form. **This model of consciousness would predict that consciousness survives bodily death (that is, when our body dies, our consciousness continues).**

Psychic phenomena and survival of bodily death are rejected by physicalist science and are often scorned as "paranormal." The word *paranormal* itself is problematic because it assumes that we know what "normal" is. It's conceivable that what we consider to be paranormal or impossible today only *seems* that way because of our primitive understanding of reality. After all, wireless cellphone technology would have seemed paranormal to people two thousand years ago.

Evidence

In *An End to Upside Down Thinking*, I evaluated a number of such paranormal phenomena ("anomalies") and argued that if even *one* of them is real, then physicalism is in big trouble. On the other hand, we can explain the phenomena much more easily by regarding consciousness as the basis of reality. And, in fact, there's a ton of credible evidence in more than one category—in spite of loud attempts from skeptics to discredit the evidence on Wikipedia and other media outlets. This taboo has forced many interested in studying these topics to leave traditional academia. Thanks to brave scientists who have risked their reputations, the evidence base has accumulated to the point where someone like me can feel comfortable writing books about it.

The reality of these phenomena points toward a view of life that is comforting. I used to reason that comforting views of reality were likely rationalizations that people just made up. As physicalist physicist Lawrence Krauss summarized it: "We all *want* there to be more out there."[20] But just because something is comforting doesn't mean it's false. It's possible for something to be both comforting and true at the same time. And as we'll see, that's what our science suggests.

Next, I will summarize evidence from multiple areas which, *independently*, converge toward the One Mind theory. The four areas we will review are:

1. Consciousness "anomalies"
2. Direct experiences
3. Physics
4. Philosophy

Consciousness Anomalies

Psychic phenomena

A number of scientists over many decades have tested the following phenomena scientifically. The effects are typically small but highly statistically significant.

○ *Remote viewing:* The ability to perceive something far away in space and time with the mind alone

- Example: Starting in the 1970s, the US government ran a psychic spying program (using remote viewing for national security) for more than 20 years, spending roughly $25 million. The program was run out of the Stanford Research Institute. Recently declassified CIA documents explicitly state: "Remote viewing is a real phenomenon[;]... Implications are revolutionary." Former president Jimmy Carter also confirmed that remote viewing was used to find a missing plane in an African jungle that radar systems were unable to locate.[21]

○ *Telepathy:* Mind-to-mind communication

- Example: In the "ganzfeld" experiment, one participant (the "sender") is shown an image and asked to telepathically send the image to a person in another room ("the receiver"). The receiver isn't shown the image. The experimenters later present the receiver with four pictures. They ask the receiver to pick which one the sender is sending "telepathically." If there is no effect, the receiver should guess correctly 25 percent of the time (one out of four). However, over many trials, and over many decades, the receiver guesses correctly closer to 32 percent of the time. This is massively significant from a statistical standpoint.[22]

○ *Precognition:* Knowing or sensing the future before it happens

- Example: Participants sit in front of a computer screen while their physiology is being measured (skin, brain, heart, eyes). The computer screen randomly generates neutral images or arousing ones. Participants' physiology spikes just *before* the arousing images appear, without consciously knowing what kind of image they will see. No one consciously knows what image will appear because the pictures are randomly generated. This study suggests that the body subtly senses the future before the future is known.[23]

○ *Psychokinesis:* Mind impacting matter

- Example: Random number generators (RNGs) are machines that randomly generate 1s and 0s. About 50 percent of the time the machines generate 1s, and about 50 percent of the time they generate 0s. It's sort of like flipping a coin over and over. Participants are asked to mentally "make" the machine produce more 1s than 0s, without making physical contact. Participants *are* able to make the machines produce *slightly* more 1s than 0s, statistically speaking. RNGs are also set up all over the world and have behaved nonrandomly when major global events occur (for example, 9/11). These studies suggest that there is an influence of collective consciousness on physical systems, even though most people don't know the machines are running. The implication is that large groups of people impact physical reality when they collectively focus in the same direction. And they don't have to be aware that they're having an impact in order to have an impact.[24]

A few highlights:

○ *Six sigma results:* The aforementioned psychic phenomena have achieved "six sigma" statistical results under controlled conditions. This means that the odds the effects were due to chance are more than a *billion to one.* In other words, psychic phenomena are statistically real effects.[25]

○ *Leading statistician:* Statistician Dr. Jessica Utts, the 2016 president of the American Statistical Association, stated in her 1995 report to the CIA and US Congress: "Using the standards applied to any other area of science, it is concluded that psychic functioning has been well established."[26]

○ *Mainstream academic journal meta-analysis:* Lund University psychologist Dr. Etzel Cardeña ran an analysis of the accumulated studies on these phenomena. His paper was published in 2018 in *American Psychologist*, the official peer-reviewed academic journal of the American Psychological Association (that is, a mainstream journal). As summarized by Cardeña: "The evidence provides cumulative support for the reality of [psychic phenomena], which cannot be readily explained away by the quality of the studies, fraud, selective reporting, experimental or analytical incompetence, or other frequent criticisms. The evidence... is comparable to that for established phenomena in psychology and other disciplines."[27]

○ *Princeton's Engineering Anomalies Research Lab:* Such phenomena were also tested and scientifically validated at Princeton University in a lab run from 1979 to 2007 by Princeton's former dean of engineering, Dr. Robert Jahn, and laboratory manager Brenda Dunne. As Jahn told the *New York Times* when he closed the lab, "If people don't believe us after all the results we've produced, then they never will."[28]

○ *Animals:* The work of former Cambridge University biochemist Dr. Rupert Sheldrake suggests that animals have psychic abilities as well. For instance, he conducted studies on dogs that know when their owners are coming home. He found that there was a real statistical effect under controlled experimental conditions (sending the owner miles away from home in a car that wasn't hers and sending her back home at randomly selected times): the dog waited by the window when the owner was coming home.[29]

Survival of bodily death

○ *Near-death experiences (NDEs):* While a person is in a
state of little or no brain function, he or she has a highly
lucid consciousness. The following are typical stages,
although not everyone experiences (or remembers) all of
them: positive emotions (some NDEs are distressing,
however); heightened senses; acknowledgment of being
dead; out-of-body experiences; encountering heavenly
realms; experiencing a dark space or tunnel; encountering a brilliant light, beings of light, or other mystical
beings and deceased relatives; a sense that space or time
is different; a panoramic life review; learning special
knowledge; a "flash-forward" to the future; and perceiving a boundary and returning to the body (either
voluntarily or involuntarily). Sometimes these experiences occur in people who've been blind since birth, yet
they report being able to see during the NDE. Phenomena resembling what we now call NDEs have been
reported for thousands of years (including in *The Tibetan
Book of the Dead, The Egyptian Book of the Dead*, and Plato's *Dialogues*).

Physicalists have tried to claim that NDEs are simply
hallucinations caused by a dying brain. One of the biggest pieces of evidence refuting that hypothesis (among
others I describe elsewhere) is the "veridical out-of-body
experience." These are instances in which a person's
consciousness leaves his or her body and has visual perceptions, which, upon being resuscitated, are *verified* as
accurate by others. **Accurate memories are, by definition, not hallucinations**. In some cases, these verified
memories are time-stamped as occurring during the
time when the brain should not have been able to produce any memories—let alone such vivid, accurate ones.
This poses a serious problem for physicalism.

The book *The Self Does Not Die: Verified Paranormal
Phenomena from Near-Death Experiences* (2016) by

Titus Rivas et al. includes more than a hundred such well-documented accounts. For example, one woman experienced her own resuscitation from outside her body and also accurately saw what was happening in the operating room next door. Not even her own doctor had known what was happening in that room.[30] According to the physicalist view of consciousness, instances such as this should *never* occur. Yet there are many cases. **The implication is that consciousness exists independently of a functioning brain.**[31]

○ *Mediumship and after-death communications:* Throughout history there have been cases of extraordinary individuals (known as "mediums") who seem to have an ability to communicate with the deceased. In addition to strong anecdotal cases, the Windbridge Research Center has recently run controlled studies to test mediums using *five levels of blinding.* The results, as described in peer-reviewed journal articles, suggest that some mediums can in fact obtain information about dead people that can't be explained by ordinary means. There are also many anecdotal accounts of more spontaneous encounters with the deceased, some of which are veridical (accurate information is obtained).[32]

○ *Children with past-life memories:* Since the late 1960s, the University of Virginia's Division of Perceptual Studies has examined more than 2,500 cases of young children who describe memories of a life that is not their own. The children are typically between the ages of two and five. In the most compelling cases, the researchers have been able to find historical records that align with what the child describes. In other cases, the child has a birthmark or physical defect that aligns with how the person died in the previous life. Sometimes these cases are validated by medical records of the deceased person. It's as if, when a whirlpool goes back into the broader stream, the water is "recycled" into a new whirlpool that carries information from the previous one.[33]

Many eminent scientists have examined these phenomena and have either concluded that they're real or they've taken a serious interest in them. Cardeña has compiled a comprehensive list, with associated sources, at the "Psi Encyclopedia" website.[34] Listed here are examples of Nobel Prize winners who met Cardeña's criteria (parenthetically, I've included the year they won the Nobel Prize and the category): Brian Josephson (1973, physics); Marie Curie (1903, physics, and 1911, chemistry); Pierre Curie (1903, physics); Max Planck (1918, physics); Kary Mullis (1993, chemistry); Jean Perrin (1926, physics); Charles Richet (1913, physiology or medicine); Glenn Seaborg (1951, chemistry); Eugene Wigner (1963, physics); and JJ Thompson (1906, physics).

Summary: There is strong scientific evidence to show that humans have psychic abilities and that our consciousness continues after we die—and this has been studied by many brilliant people. No one has come close to being able to shoot down *all* of the evidence. If even *one* instance is credible, we have to seriously question physicalism.

Direct Experiences

The notion of the One Mind can be difficult to process because our ordinary experience is one of separation. Our perceptual systems (eyes, ears, and so on) cause us to interpret a "me" here and a "you" there.

In everyday states of consciousness, separation is a reality, but in "mystical" states, people claim to experience deep interconnectivity directly. Such states include:

- Near-death experiences

- Psychedelic trips

- Other "spiritual" experiences induced by practices such as meditation and beyond (discussed further in chapter 5)

In these cases, the brain's "filter" appears to be unlocked, and people experience the broader reality that is ordinarily obfuscated. So the unity has always been there, but the brain has shielded us from it.

Before going further, I need to mention a critical point to keep in mind in this section and throughout the book: **Studying mystical experiences as a third party is challenging because we're talking about someone else's subjective experience.** If we haven't experienced something ourselves, it can be difficult, if not impossible, to relate to. Let's say you tried to explain the taste of chocolate to someone who's never tasted a dessert. Words just wouldn't suffice. Human language is inherently limited; it's an approximation but not a substitute for the real thing. Tennis champion Roger Federer could write a book on how amazing it feels to win Wimbledon, but until we do so ourselves, we can't know exactly what it's like.

Reports of mystical states are anecdotal, which some critique as being less scientific than other modes of investigation. Anecdotal evidence is certainly less convincing when we have just a few anomalous cases. But when there are *many* similar accounts suggestive of something universal, we have no choice but to pay attention. If a patient walks into a doctor's office saying, "I have a headache," should the doctor say, "Hmm, that's anecdotal, I can't help you"? Of course not. We can be scientific while still paying careful attention to anecdotal reports, as long as we maintain an appropriately critical and skeptical eye.

Descriptions of mystical states

While not all mystical states are identical, they have many similar characteristics that converge toward the One Mind view of reality. Psychedelics researcher Rick Strassman, MD, summarized the phenomenon well: "There no longer is any separation between the self and what is not the self. Personal identity and all of existence become one and the same. In fact, there is no 'personal' identity because we understand at the most basic level the underlying unity and interdependence of all existence. Past, present, and future merge together into a timeless moment, the now of eternity."[35]

Think about how incredible that is! People in mystical states feel at one with everything and don't experience time? If we've never experienced that ourselves, how can we relate? Therefore, it's easy for us to not let this sink in.

Dr. Bonnie Greenwell has counseled thousands of individuals who have gone through spiritual experiences along these lines (known as "awakening experiences," which we'll discuss in chapter 5). She quotes one such person who gives a classic description: "I was hiking on an island and looked out across the water to nearby land. Suddenly I saw that everything inside me was outside me, and everything outside was inside. There was no separation. The water, the hills, trees, earth were in me—and my bones, blood, flesh, organs were in them. We were all one."[36]

Perhaps the best experiential evidence of the One Mind comes from the life-review phenomenon sometimes reported during near-death experiences. People relive their lives *through the eyes of people they impacted*. Sometimes they even feel how a third party was impacted by an action. This phenomenon is the most impactful topic I've studied. As one near-death experiencer stated:

> All of my life up till the present seemed to be placed before me in a kind of panoramic, three-dimensional review, and each event seemed to be accompanied by a consciousness of good or evil or with an insight into cause or effect. Not only did I perceive everything from my own viewpoint, but I also knew the thoughts of everyone involved in the event, as if I had their thoughts within me. This meant that I perceived not only what I had done or thought, but even in what way it had influenced others, **as if I saw things with all-seeing eyes**. And so even your thoughts are apparently not wiped out. And all the time during the review the importance of love was emphasized. Looking back, I cannot say how long this life review and life insight lasted, it may have been long, for every subject came up, but at the same time it seemed just a fraction of a second, because I perceived it all at the same moment. Time and distance

> seemed not to exist. I was in all places at the same time, and sometimes my attention was drawn to something, and then I would be present there.[37] [emphasis added]

It's as if, in this alternate dimension of reality, consciousness is liberated. And we're able to see through the eyes of the One Mind via multiple lenses, whereas while in a body, we only have access to our whirlpool. **The life review is very direct experiential evidence for the One Mind.**

We have to wonder: Do mystical experiences explain why spiritual traditions all over the world have such a similar, One Mind-esque view of reality? Whether its mysticism from the East (aspects of Buddhism and Vedanta in Hinduism) or even the West (Gnosticism in Christianity, Kabbalah in Judaism, and Sufism in Islam), we find a similar worldview around a unified, interconnected consciousness. Perhaps the similarities arise because the *experience* of reality is universal. Each culture has individuals who have experienced general features of the same reality, but they might have used different words to describe it. And then those words were translated by people who hadn't directly had the experiences themselves. We can easily imagine how ideas could become distorted or lost in translation.

Summary: There are many mystical experiences that align with the One Mind view of reality. It's not just a hypothetical, intellectual idea; it's a reality often felt—*and it is universal.*

Physics

Physicalism purports that physical matter is ultimately what creates consciousness. But what *is* matter? It's solid stuff, right? And the universe is just the interactions of those solid pieces of matter, sort of like billiard balls. Unfortunately for physicalism, science is teaching us that this perspective is outdated.[38]

If we look at atoms under a microscope, we find that they are 99.99999999 percent composed of *empty* space. Furthermore, the emerging field of quantum physics, originally conceived in the early 1900s, has introduced an additional wrinkle. In the famous double-slit experiment, it was discovered that an allegedly solid particle only acts like a solid particle if it's being observed. When it's not being observed, it behaves like a wave of probability (it's maybe here, maybe there). In essence, matter doesn't act like something solid all the time.

So, allegedly "solid" matter is mostly empty space and doesn't even act like solid matter unless it's being observed. As physicist Hans Peter Dürr said, "Matter is not made of matter."[39] And this is the stuff that physicalists say gives rise to consciousness? The truth is that we don't even know what matter is. This seems problematic for a philosophy using that very thing as the foundational element of its reality. Nobel Prize–winning physicist Werner Heisenberg summarized the mind-boggling implications well: "Quantum theory does not allow a completely objective description of nature."[40]

Quantum physics also points us toward the notion of interconnectivity through the phenomenon of "entanglement." The basic finding is if you impact a particle located in one place, an entangled particle far away reacts in a correlated manner *at the same instant*. Entanglement suggests that there's an invisible connection between distant particles—no matter how far apart they are in space and time—which we can't see with our eyes. Albert Einstein famously called this "spooky action at a distance" and tried to disprove it. Ironically, his exploration into the topic only led to his further proving the reality of entanglement.

Summary: Quantum physics provides a major challenge to physicalism, while supporting a consciousness-based, interconnected reality.

Philosophy

The One Mind view of reality—which is akin to metaphysical *idealism*—is also philosophically superior to physicalism. In other words, even if we ignore all of the science we've discussed, physicalism runs into serious philosophical problems.

I will be presenting a simplified overview here, but I give you

fair warning that it's still abstract and heady. At the end of the section, there is a short summary, so if you find this confusing, then feel free to skip the meat of the section. This concept took me a while to wrap my head around. The main point is an important part of my overall argument; therefore, I felt obliged to include it despite its more challenging nature.

If, on the other hand, you find this summary lacking in detail (I watered it down as much as possible), I recommend reading Bernardo Kastrup's books, particularly *The Idea of the World* (2019). He offers a full-on defense with peer-reviewed philosophical essays, whereas here I'm merely outlining his basic argument.

Philosophers use the principle of Occam's Razor—"the simplest solution is usually the best"—to evaluate the merits of an idea. If a solution follows Occam's Razor, it makes no more assumptions than are necessary. Such a solution is said to be "parsimonious." The claim of idealism (that is, the One Mind) is that it is more parsimonious than physicalism: It can explain reality at least as well as physicalism while relying on fewer assumptions.[41] And therefore it should be philosophically preferred over physicalism.

As Bernardo Kastrup argues, in order for the physicalist idea that "matter (via the brain) creates consciousness" to be true, the following four statements about reality *all* must be true:[42]

1. Your conscious perceptions exist.
2. The conscious perceptions of other living entities, different from your own, also exist [that is, other people are conscious].
3. There are things that exist independently of, and outside, conscious perception [for example, there was a universe that existed before consciousness evolved].
4. Things that exist independently of, and outside, conscious perception generate conscious perception [for example, the universe is what gave rise to consciousness (via the evolution of a brain)].

#1 is obviously true. You are conscious as you read these words right now. In fact, your own consciousness is the *only* thing we can 100 percent prove. *Everything else in life is an inference.*

It's worth noting, however, that some philosophers have tried to claim that consciousness is merely an illusion. Tufts professor Dr. Daniel Dennett is one such philosopher. But if he were to say, for instance, "I think consciousness is an illusion," we'd have to ask him what the "I" is in his sentence. It seems as if any of his own statements would immediately suggest that *he* is the consciousness speaking. I disagree with the idea that consciousness doesn't exist by virtue of the fact that I am undeniably having an experience right now as I type these words. If consciousness is in fact the basis of all reality, as I'm suggesting, then think about how absurd it is to say there's no such thing as consciousness!

#2 asks us to go a step further and assume that other people are also conscious. This feels like it's true, but we can't prove it. I know what it's like to be conscious. Others around me act in a similar way. So while I can't prove they are also conscious, I have no evidence to suggest that they're all just zombies faking it. Therefore, it seems fair to take a small leap of faith and infer that humans around me are also conscious. That seems like a reasonable step given that I already know what consciousness feels like myself.

If we accept just #1 (I alone am conscious), we are left with what's known as *solipsism*. If we accept #1 and #2 together (I am conscious and so are other people), we've met the requirements of idealism/ the One Mind. Consciousness is all there is, and it exists within me and others. And if all reality is indeed just one universal consciousness—one stream in which each of us is a whirlpool—that would explain why we all feel like we inhabit the same universe— not because a universe exists outside of us, but because we all exist within the same stream. Most of us can probably get behind the reality of both #1 and #2.

So far, we haven't needed to introduce the concept of a brain or a universe outside consciousness. While we might be tempted to assume that #1 and #2 are true because the brain creates my consciousness and others' consciousness, we should remember that this is precisely the assumption we are questioning. All we're saying so far is that consciousness exists, without using preconceived notions to explain how it got there.

Next, things get dicey, as #3 asserts that something exists independently of conscious experience. It's sort of like the often-asked question: "If a tree fell in the woods and no one was there to hear it, did it fall?" I'm actually pointing to an even more basic question: "If there were no consciousness anywhere in the universe (in a body or otherwise), could we prove that the tree exists?" The answer is so obvious that it's almost difficult to comprehend. If there were no consciousness to register the experience of the tree, of course, by definition, we couldn't prove that it exists. In order to prove its existence, we would need a consciousness to experience the tree (for example, seeing, touching, or smelling the tree).

This brings us to a very basic fact of life: *In order to verify that anything exists, by definition, we need consciousness to experience it.* To say there exists something outside the experience of all forms of consciousness is simply unverifiable.

So if we wanted to be true skeptics, we would have to acknowledge that we can't prove there was a universe existing before consciousness allegedly "emerged via evolution." You might ask, "What about the light we can measure from the early days of the universe?" The fact that we now see light traveling from the universe's past tells us nothing about whether consciousness existed back when the light was formed.

So believing that a world exists outside of consciousness would require a significant leap of faith. We had to take a small leap of faith with #2, but at least that leap of faith was based on something we directly experience (consciousness). We *know* the category of conscious experience. #3 is asking us to take a leap of faith on something that is by definition not experienceable. It is asking us to conjure up an entirely new category (known in philosophy as an "ontological category").

#4 makes an even bigger claim. It takes this unprovable category—the unverifiable world existing independently of consciousness (#3)—and says *that very thing* is responsible for producing conscious experience (via the evolution of a brain). It asserts that the unprovable unknown gives rise to the one thing we do know for sure: consciousness.

Now, this doesn't disprove physicalism nor does it prove the One Mind. But if we're following Occam's Razor, then we have to ask ourselves: Why do we need to incorporate #3 and #4 when we can explain reality with #1 and #2 alone? This is where physicalism runs into a huge philosophical dilemma, notwithstanding all of the scientific problems. **Physicalism is problematic from a philosophical standpoint because it makes unnecessary assumptions and therefore lacks parsimony. It relies on a leap of faith that we don't have to take in order to explain reality.**[43]

Ironically, physicalists (some of whom claim to be skeptics) base their entire worldview on this leap of faith. Einstein believed in a world outside consciousness, although he humbly admitted: "I cannot prove that my conception is right, but that is my religion."[44]

I'd argue that physicalism is much more faith based than the alternative I'm proposing. Physicalism is a huge gamble, a spin of the roulette wheel. My view is that it's not a wise bet.

Summary: Using the philosophical principle of Occam's Razor—"the simplest solution is usually the best"—the One Mind model is superior to physicalism.

The Significance of Physicalism's Collapse

The dominant mainstream view of reality is clearly crumbling. Physicalism makes the assumption that the brain *creates* consciousness with no supporting evidence beyond correlation. On the contrary, when examining consciousness "anomalies," mystical experiences, physics, and philosophy, we find very clearly that the One Mind paradigm gets us much closer to explaining reality. Is the One Mind paradigm exactly 100 percent "it"? I really do not know, nor could I prove it. All I can prove is consciousness—it feels like I'm conscious right now.

My perspective on the One Mind is that it's beyond all human conceptualization. To even call it something with language is to artificially put a limitation on something inherently limitless. The One Mind framework is a *framework* alone. We need something as a basis for communication. As new data points come in, this view could easily become refined in the future. But based on all of the evidence I've seen, the One Mind view is the best directional framework we can use to describe reality. The more I research, the more the evidence converges in this direction. So that's what we'll be using throughout this book.

Why isn't this mainstream yet?

The unfortunate reality is that in spite of the mountain of evidence, the topic is considered "taboo." Psychologist Imants Barušs and cognitive neuroscientist Julia Mossbridge elaborate on the sad truth we face today:

> As a result of studying anomalous phenomena or challenging [physicalism], scientists may have been ridiculed for doing their work, been prohibited from supervising student theses, been unable to obtain funding from traditional funding sources, been unable to get papers published in mainstream journals, had their teachings censored, been barred from promotions, and been threatened with removal of tenured positions. Students have reported being afraid to be associated with research into anomalous phenomena for fear of jeopardizing their academic careers. Other students have reported explicit reprisals for questioning [physicalism], and so on.[45]

Worse than that, many scientists won't even look at the evidence. It's similar to what Galileo faced when members of the clergy wouldn't look in his telescope to see the evidence that Earth revolved around the sun. We'd think "science" would want to explore new frontiers. But instead, ironically, science is turning into a new fundamentalist religion. And compared to Galileo's era, the consequences here are far more severe for our species.

Hopefully the scientific community will take off its blinders so that our education system is given room to shift, along with societal thinking.

Many of the scientists who are brave enough to study these phenomena tell me that the mainstream is secretly interested. They sometimes receive private emails from mainstream scientists saying, "I'm fascinated by your work, but I can't talk about it publicly or let my colleagues know about my interest." We can only hope that these scientists will start coming out of the closet.

Why this topic is so immensely important

The One Mind view of reality suggests that we all have innate psychic abilities and that when our body dies, our consciousness does not. Even more than that, it implies that humans are infinite beings fundamentally interconnected with everyone and everything else. This is huge!

Physicalism, on the other hand, purports that we are finite beings destined to die, and are fundamentally separate. We're "sort of" connected because we are biological beings who share similar genes as part of the same species. And we inhabit the same planet, so that's a type of a "connection." But ultimately each of us has an individual consciousness that comes from our individual brain. There is no connection between those consciousness-es, and those consciousness-es will go away when our bodies die. There is no intrinsic meaning to life.

My own personal story shows how devastating physicalism can be to an individual's life. It's not merely a harmless scientific belief system. At the end of *An End to Upside Down Thinking*, I elaborated on the broader societal implications:

> I view the belief that we are finite and separate to be the disease underlying virtually every problem in human society today. Anxiety, depression, interpersonal problems, racial and social prejudices, gender inequality, geopolitical unrest, violence, war, greed, or nearly any problem you can think of—at their core are symptoms,

> not the disease. The world's problems are caused at the most fundamental level by the pervasive underlying assumption that we are all finite, limited, and separate. And that stems from the [physicalist] belief that consciousness comes from the brain.[46]

One way to summarize this is to say that the world's problems are caused by a misunderstanding of reality. A less euphemistic way of phrasing it is to say that *ignorance* is the root problem. That's where I think we are today as a society: massively ignorant. I was ignorant myself—not because I wanted to be, but because I had no exposure to these alternative perspectives and scientific findings.

It's as if we're collectively in a zombie state, living life without any regard for what's actually going on. To use the Vedic/Hindu analogy, we think reality is a snake, when in fact it's been a rope the entire time.

Now that the veil of ignorance has become permeable, it's time to examine the rope. We'll be looking more deeply at what the One Mind perspective implies about reality and our place in it. These perspectives will serve as the foundation for setting our life's compass.

CHAPTER 3

INFERENCES ABOUT REALITY

Introduction to Inferences

Using the One Mind as our baseline framework for reality, there are many inferences we can draw about life and how we fit in. The inferences themselves are not bulletproof, or guaranteed fact. Based on everything I've seen, I regard all of them as reasonable hypotheses. I feel it's likely that these inferences are at least directionally accurate.

The inferences we will explore are as follows:

- Only the One Mind is conscious.

- Our ultimate identity is not our body.

- The One Mind could be likened to "God."

- The One Mind is intelligent.

- Randomness is an illusion.

- The One Mind has free will.

- The Golden Rule is built into the structure of reality.

○ The One Mind is made of love.

○ We all have amnesia.

○ Reincarnation is the engine of evolution.

○ Reincarnation is structured.

○ Karma is real.

Before we get into the rationale for those inferences, we need to ensure we have the right mind-set going in. So we'll begin with an exploration of two essential topics: radical humility and paradox.

Radical Humility

Caltech physicist Dr. Sean Carroll mentions a belief probably held by many other brilliant scientists and everyday people: "Mysteries abound, but there's no reason to worry... that any of them are unsolvable."[1] On the surface, this might sound harmless. But it's actually an incredibly arrogant statement, emblematic of a societal hubris that needs to be eradicated in order for us to advance.

Why should we assume that our limited brain is capable of understanding all the mysteries of life? Is it possible that certain things are simply beyond human comprehension? For example, consider the concept of *infinity*. We know that it exists, but our minds can't actually grasp it.

Evolution has conditioned our bodies to be able to perceive what biologist Dr. Richard Dawkins calls the "Middle World."[2] Our machinery hasn't evolved to be able to comprehend the massive scales of the cosmos or the micro-scales of the subatomic world. We perceive somewhere in the middle. For example, our eyes can only see a tiny fraction of light on the electromagnetic spectrum. We can only hear a limited band of sounds. The list could go on and on.

Furthermore, UC Irvine cognitive scientist Dr. Donald Hoffman's research suggests that more evolutionarily fit organisms are ones that *don't* see reality as it is. And that allows them to survive better than if they saw reality exactly as it is. We don't need to see reality as it is in order to survive. He stated: "So that means I can't trust that my perceptions of space and time and physical objects are an insight into the nature of reality.... Our perceptual systems are a species-specific adaptation not designed to show reality as it is, but in fact shaped to hide reality, because we don't need to know reality; it's unnecessary."[3]

In other words, we're working with equipment that's naturally incapable of certain types of comprehension. And yet we are using that machinery to try to understand reality. **Would it be possible to teach an amoeba multivariable calculus? We could easily be like the amoeba.**

Here's another analogy from the satirical novella *Flatland* (1884) by Edwin A. Abbott. Imagine there exists a society of people living in a two-dimensional world (2D Land). It would be like living in a flat piece of paper. Now imagine that a sphere (three-dimensional, or 3D) intersects with 2D Land. When the sphere first touches 2D Land, the intersection manifests in 2D Land as a single point. Everyone in 2D says, "Look, it's a point!" No... it's actually a sphere. The people in 2D Land, by design, simply cannot perceive that it's a sphere. All they can perceive is the intersection point. Their interpretation is way off. Now, imagine the sphere dips down even further. It then intersects 2D Land with the appearance of a circle. The people in 2D Land say, "Look, it's a circle!" No... it's actually a sphere [image follows].

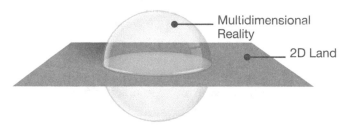

People in 2D Land see a circle. In reality, it's a sphere.

Now, imagine an even more complex example. What if a 3D horseshoe were to intersect with 2D Land? The people in 2D Land would say, "Look it's two separate circles!" No... it's actually one contiguous 3D horseshoe [image follows].

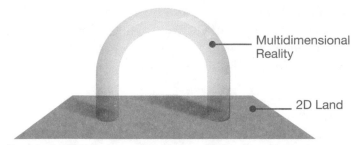

Multidimensional
Reality

2D Land

People in 2D Land see two separate circles. In reality, it's a
3D horseshoe.

My point is: we need to be open to the possibility that we are like
the people in 2D Land trying to extrapolate conclusions that are
dead wrong because we are literally incapable of perceiving fully.

Science acknowledges how much of life is a mystery. For instance,
only roughly 4 percent of the universe is made of known matter,
whereas the remaining 96 percent is comprised of "dark matter"
and "dark energy." So we don't know what the majority of the
universe is. And we don't even have a unifying theory of physics
that brings all of our individual theories together. That means our
supposed "laws" of physics aren't telling us the full picture. We
should expect something radical to come along. We actually *need*
it in order to develop a more comprehensive theory of reality.

Furthermore, the only thing we can truly know with absolute
certainty is: "I have the feeling of being conscious at this very
moment as I write these words." Everything else—*everything*
else—is an inference of some kind.

Let's examine some very basic "facts" we probably don't even realize are inferences. For example, consider "time" and "space." We
seem to think we know what they are. The clock ticks in a predict-
able manner, and I can measure the distance between two keys on
my keyboard.

But time and space aren't so simple in reality. Einstein's relativ-
ity theory teaches us that time and space are different depending

on conditions such as speed and gravitational force. They aren't fixed. Time can speed up or slow down, and space can contract or expand.

Our own introspection also shows that time and space are strange. When I think back to an event in the alleged past, I realize that the event occurs as a memory. The memory is a thought in my mind. And that thought occurs right *now*! There is no past we can validate; there is only the memory of it, which occurs in the present moment. If we watch a movie from the past, that watching occurs *now*. The future is similarly funky. It occurs now as a thought in the present moment. The idea of linear time is a perceptual interpretation; it is not an experiential reality. Space is similarly elusive. We perceive objects that we interpret as being "over there," but in reality those objects are appearing within our conscious experience, occurring *here*. No existence can be proven beyond our own consciousness *here and now*.

And if there is only *here and now*, then how could it be possible that "causality" exists—the notion that something from the past causes something in the present and future? In the *here and now*, nothing is causing anything. Everything simply *is*. Everything is simply a spontaneous emergence.

When I think of the word *humility*, it conjures up images of not bragging about skills. For example, a humble tennis player wouldn't brag about how talented she is, and a humble CEO wouldn't brag about how much money she made last year. Throughout the remainder of this book, I'll be referring to a much broader, more radical humility: the notion that the human mind is simply incapable of comprehending reality. And furthermore, I'd argue that the extent of our "not knowing" is much more than we tend to think.

I can say from personal experience that the more I study the nature of reality, the less I feel like I know. Former US defense secretary Donald Rumsfeld famously referred to "known unknowns" and "unknown unknowns." The more I learn, the more "known unknowns" appear. And I assume that the extent of "unknown unknowns" is beyond all comprehension. I constantly feel like a beginner (in Zen Buddhism known as *shoshin* or *beginner's mind*).

Ironically, it is through the building of our intellect that it becomes capable of recognizing its limitations. At that point, the brutally honest intellect has no choice but to self-destruct. It's like a calculator that can't compute high-enough numbers. We are thus left in a place of radical humility.[4]

It is from this place of not knowing that we will explore ideas that might sound far-fetched at first glance. On top of that, we're using the limited tool of language to examine these ideas. But if we remember how little we know, we will remain open to possibilities. That's not to say we should believe everything blindly. Rather, I'm arguing that we need to be open... open to everything.

Paradox

A paradox is a seemingly self-contradictory idea. Our linear, human mind prefers to think in black and white—this *or* that. Paradox doesn't do well with that mind-set because it instead requires a "both/and" perspective ("both this and that" are true).

I've come to realize that until we can hold seemingly paradoxical ideas in unity, we're not going to get anywhere. This is *essential*. For this reason, *Buddha at the Gas Pump* podcast host Rick Archer has often said: "I'd like to have a T-shirt made that has the word *Paradox* on one side and *Ambiguity* on the other." (I'd probably add the word *Humility* to both sides of the T-shirt.)

This harkens back to the Buddhist notion of the "middle way": we are best served by not getting rigidly stuck in a single point of view. Reality is multifaceted, in spite of our human desire to keep things simple.

Our ability to hold a paradox comes from an ability to zoom our perspective in and out. If we assume reality is the One Mind, then if we zoom our lens all the way out to that highest level of reality, there is no separation, no space, no time, and no individuals. There is only the single infinite field of unified awareness. This view is often called the "Absolute" perspective. (We'll stick with that term because it's so commonly used. However, the term *Absolute* is perhaps misleading because it implies a stagnancy to the highest-level state, when, in fact, the One Mind seems to be dynamic.)

To tie this to common spiritual parlance, we could equate the Absolute with *nonduality* ("not two") and the Relative with *duality*, or "the individual ego."

From the Absolute, nondual perspective, there is no universe, no people, and nothing ever happened. It's all just an infinite ground of all being, beyond space and time, without any separation.

This Absolute view makes no sense from the perspective of our personal experience. We feel like an individual, living in a universe with space and time, seeing events from the past causing other events, feeling a range of emotions, and so on. From the lens of a human being, a whole lot is happening. So if we zoom from the Absolute to the Relative, we experience a very different reality.[5]

The paradox here is that *both* the Absolute and Relative are true at the same time. If the One Mind paradigm is our meta-paradigm of reality, then holding the Absolute with the Relative could be considered the "meta-paradox."

Archer often references an analogy from the Vedic tradition to explain this notion. Imagine a room with a bunch of clay pots. At one level, there are no pots; there is only clay. At another level, there *are* pots. The Absolute and Relative are simultaneously true.

As we delve into admittedly heady, abstract topics, we'll be referring back to this terminology. With this in mind, we'll now move on to inferences about the One Mind reality.

Only the One Mind Is Conscious

The philosophy of "panpsychism" is starting to gain public traction. For example, recent books by Annaka Harris, Philip Goff, and Christof Koch have given the idea significant airtime. Panpsychism rightly acknowledges the challenges with physicalism (that is, the unproven belief that the brain creates consciousness). But it makes a wrong turn: it posits that consciousness is an inherent property of matter. Each unit of matter naturally has consciousness in it. When you combine units of matter, you get more complex consciousness. Panpsychists argue that everything is conscious, including animals, plants, and even rocks. (More specifically, this is known as "constitutive panpsychism." While there are other forms of panpsychism, people often mean "constitutive panpsychism" when they say "panpsychism." So for the majority of this book, I will simply call it panpsychism without the "constitutive" qualifier.)

This is *not* what I'm arguing. In my view, panpsychism takes a step in the right direction, but it's still wrong. It's really just another form of physicalism. It starts with matter and says consciousness comes *from* it. The diagrams that follow illustrate this idea.

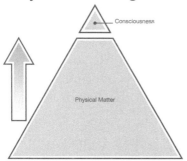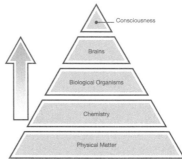

(Constitutive) Panpsychism: Consciousness is a property of matter.

Physicalism: Matter (via a material brain) creates consciousness.

Because panpsychism is still just a form of physicalism, it introduces no intrinsic meaning into life. For this reason, I consider it very dangerous (as do others, such as Bernardo Kastrup[6]). Panpsychist Philip Goff asks: "What makes the case that intelligent life is a thing of value?"[7] This statement might sound horrible, but he's asking the right question if you buy into his panpsychist view. Matter has consciousness, but "meaning" still doesn't exist anywhere.

Now let's get into why this theory is *not* what I'm arguing.

An underlying reason for the movement toward panpsychism is perhaps due to Thomas Nagel's popular definition of consciousness from the 1970s: "Consciousness is what it's like to be something." In other words, consciousness is what it's like to be a human or a dog or anything that has consciousness.

At first glance, this seems reasonable. But as noted by philosophers such as Rupert Spira, embedded in Nagel's definition is a big, unwarranted assumption. It assumes that there exists a human that could be conscious—that there is a human *from which* consciousness arises. See the issue there? He's assuming physicalism in his definition.

On the contrary, the One Mind model suggests that all reality is just one universal consciousness. There is no human or dog or insect or plant or stone that could be conscious. Rather, as Spira says, "Only consciousness is conscious."[8] And that one consciousness—the One Mind—experiences an apparently physical world *through the vehicle* of humans and other physical vessels.

The One Mind is the infinite, silent context underlying all experience—the field of reality and the ground of all being. Spira often compares it to an infinite screen. The screen's pixels are colored when a movie is playing and various characters appear (for example, humans). But no matter what happens to the characters on the screen and no matter what colors light up the pixels, the screen is untouched.

The One Mind is the substrate of all experience. The apparent diversity in the world is a reflection of "modulations" of consciousness[9]—like waves appearing in an ocean.

Our Ultimate Identity Is Not Our Body

What are we? This answer seems obvious. Our everyday experience suggests to us that our identity is our body. That's what our perceptions show us. It's also how our mind interprets reality and certainly what modern society teaches us. I'm a human being named Mark Gober. Easy.

But based on the One Mind model, our identity at the Absolute level is not our body or our personality; rather, it's our consciousness—the one consciousness at the core of all existence. Our identity is the full "stream" itself, the One Mind. The body is just something that consciousness experiences. We *inhabit* the body, but we aren't *identified* as the body.

That said, at the Relative level of reality, it sure feels as if we are a body and a personality. That's the way we live day to day. So in order to buy into the One Mind model, we have to hold the paradox of feeling like an individual in the context of being the totality of existence. We are both a whirlpool *and* the stream simultaneously.

That is one of the most earth-shattering implications of the One Mind model: we've been thinking about our very own identity incorrectly! As is often stated: "We aren't human beings having a spiritual experience; we are spiritual beings having a human experience." Or alternatively, as author Suzanne Segal phrased it: "We are sense organs of the Infinite."[10]

The shift away from physicalism toward the One Mind is likely a bigger revolution in thinking than any other in human history. The realizations that Earth isn't flat and that Earth isn't at the center of the solar system were enormous paradigm shifts, but they pertained to how we viewed the external world around us while preserving our identity. What I'm talking about here is shifting how we think about who and what we fundamentally are.

The One Mind Could Be Likened to "God"

God is a term with lots of baggage. The Judeo-Christian cultural view anthropomorphizes God. God is often depicted as a white, bearded male in the sky who exists separately from us; and who often exhibits vengeful, judgmental, and bellicose behavior. We are here, and God is "up there" in the heavens. Obey "him"... or else. People throughout history have justified murder and mayhem in "his" name.

Debates rage on as to whether or not God exists. However, a first step that people often skip is to make certain we're using a consistent definition. Let's be sure we're talking about the same guy before we argue!

The model in this book is not claiming that there exists a separate God having the characteristics just described. Rather, I'm arguing that the One Mind—a genderless, impersonal, infinite field of consciousness—is the basis of reality. There is one infinite stream in which we are localized whirlpools. If one wanted, one might call the One Mind—the entire stream—"God." Rick Archer has often stated as a rebuttal to atheistic views: "I don't believe in the same God you don't believe in." Or as I like to say, "They got the wrong guy."

If we were to employ this One Mind definition of God, there are some important implications. Since each us is made of the same water in the same stream, then it would logically follow that each of us is an aspect of God. This aligns with the spiritual notion that God is both *immanent* (within) and *transcendent* (beyond). There is no separation between God and us. And yet there *is* separation because we don't access the full stream in daily living. Both are true. Another paradox.

Returning to our discussion about identity, we could reframe my previous comments using God-centric terminology. We could say that we are "tentacles of God" or "instruments of the Divine." This is why spiritual author Ram Dass advised: **"Treat everyone you meet like God in drag."**

Certainly in Eastern religions we see ideas like this. But they also appear in the West if we look closely. For example, consider Jesus Christ's quote: "I and the father are one." Translation: *Jesus, as an individual whirlpool, is one with the stream.* And consider his quote: "The kingdom of God is within you." Translation: *Even though you are an individual whirlpool, you are comprised of the same water that makes up the full stream of reality.* Words such as *father* and *God* have become so culturally off-putting that we can overlook what's actually being said.

In my life, the term *God* has been a turn-off. I used to immediately shut down when I heard it. Even when I began my explorations several years ago, I would become viscerally disturbed anytime someone used the word, even if I knew they weren't talking about a personal God. Now, I've come to terms with it *as long as it's being defined properly*. Too often people use it without specifying what they mean.

So why not just do away with the term *God* to avoid possible confusion? Some claim that the "Ah" sound in the word *God* creates a frequency that keeps us in alignment with the greater stream of consciousness. Why that would be the case, I'm not sure. Interestingly, when we look at religions around the world, we find that the very same "Ah" sound is found in words that represent divinity: *Allah* (Islam), *Brahman* (Hinduism), *Hashem* (Judaism), and so on.

As spiritual teacher Dr. Wayne Dyer said, "I was taught that the expression of the word or sound, which means God, brings us into contact with God. God is the basic fact of the universe, symbolized with this most natural and comprehensive of all sounds. It's no accident that the words omnipotent, omniscience, and omnipresent contain the sound of God"[11] [emphasis added].

However, regardless of whether or not that idea is true, I will generally use the term *One Mind* in this book—unless I'm making an explicit tie to religion.

The One Mind Is Intelligent

A basic fact of our individual existence as humans is that we are intelligent. The degree of intelligence can vary from person to person, but each of us has some intelligence.

If each of us is intelligent, and if we are individually part of the One Mind, it seems to follow logically that the One Mind is intelligent. Stated another way, if the whirlpools are intelligent, and the whirlpools are part of the stream, then by definition the stream has intelligence.

Let's take a further logical step. If all reality is just the One Mind, then all reality—ourselves and everything around us—is conceivably embedded with intelligence as well.

Everywhere we look in the universe—whether it's a flower, a human being, mathematics, the greater cosmos, or anything— we find immense complexity. And in light of this unthinkable complexity, it seems reasonable to assume that the One Mind has unthinkable intelligence.

So perhaps this is what religious traditions mean when they make statements such as "God is omniscient" and that we should "see God in everything." When we depersonalize "God," it all starts to make more sense.

Randomness Is an Illusion

The universe

As previously discussed, the physicalist view of reality tends to view the universe as random. However, if the universe is embedded with unthinkable intelligence, should we question the "randomness" hypothesis? I think so.

We have to ask ourselves: Why is the universe exactly balanced for life? Robert Lanza, MD (a stem-cell researcher who was named one of *TIME* magazine's top 100 most influential people in 2014), and physicist Bob Berman examine this question in their book *Biocentrism* (2009). In a chapter titled "Goldilocks's Universe," they state:

> By the late sixties, it had become clear that if the Big Bang had been just one part in a million more powerful, the cosmos would have blown outward too fast to allow stars and worlds to form. Result: no us. Even more coincidentally, the universe's four forces and all of its constants [tables follow] are just perfectly set up for atomic interactions, the existence of atoms and elements, planets, liquid water, and life. Tweak any of them and you never existed.[12]

List of Constants (as shown in *Biocentrism*)[13]

Name	Symbol	Value
Atomic Mass Unit	m_u	$1.66053873(13) \times 10^{-27}$ kg
Avogadro's Number	N_A	$6.02214199(47) \times 10^{23}$ mol^{-1}
Bohr Magneton	μ_B	$9.27400899(37) \times 10^{-24}$ J T^{-1}
Bohr Radius	a_o	$0.5291772083(19) \times 10^{-10}$ m
Boltzmann's Constant	k	$1.3806503(24) \times 10^{-23}$ J K^{-1}
Compton Wavelength	λ_c	$2.426310215(18) \times 10^{-12}$ m
Deuteron Mass	m_d	$3.34358309(26) \times 10^{-27}$ kg
Electric Constant	ε_o	$8.854187817 \times 10^{-12}$ F m^{-1}
Electron Mass	m_e	$9.10938188(72) \times 10^{-31}$ kg
Electron-Volt	eV	$1.602176462(63) \times 10^{-19}$ J
Elementary Charge	e	$1.602176462(63) \times 10^{-19}$ C
Faraday Constant	F	$9.64853415(39) \times 10^{4}$ C mol^{-1}
Fine Structure Constant	α	$7.297352533(27) \times 10^{-3}$
Hartree Energy	E_h	$4.35974381(34) \times 10^{-18}$ J
Hydrogen Ground State	$(r) = \dfrac{3a_0}{2}$	13.6057 eV
Josephson Constant	K_j	$4.83597898(19) \times 10^{14}$ Hz V^{-1}
Magnetic Constant	μ_o	$4\pi \times 10^{-7}$
Molar Gas Constant	R	$8.314472(15)$ J K^{-1} mol^{-1}
Natural Unit of Action	$\hbar$	$1.054571596(82) \times 10^{-34}$ J s
Newtonian Constant of Gravitation	G	$6.673(10) \times 10^{-11}$ m^3 kg^{-1} s^{-2}

Name	Symbol	Value
Neutron Mass	m_n	$1.67492716(13) \times 10^{-27}$ kg
Nuclear Magneton	μ_n	$5.05078317(20) \times 10^{-27}$ J T^{-1}
Planck Constant	h	$6.62606876(52) \times 10^{-34}$ J s $h = 2\pi\hbar$
Planck Length	l_p	$1.6160(12) \times 10^{-35}$ m
Planck Mass	m_p	$2.1767(16) \times 10^{-8}$ kg
Planck Time	t_p	$5.3906(40) \times 10^{-44}$ s
Proton Mass	m_p	$1.67262158(13) \times 10^{-27}$ kg
Rydberg Constant	R_H	$10\ 9.73731568549(83) \times 10^5$ m^{-1}
Stefan Boltzmann Constant	σ	$5.670400(40) \times 10^{-8}$ W m^{-2} K^{-4}
Speed of Light in Vacuum	c	2.99792458×10^8 m s^{-1}
Thompson Cross Section	σ_e	$0.665245854(15) \times 10^{-28}$ m^2
Wien Displacement Law Constant	b	$2.8977686(51) \times 10^{-3}$ m K

The universe is configured "just right" for life to exist. Somehow, life emerged from nonlife. Given what we've discussed about the One Mind, do we think the universe's fine-tuned structure is a function of the One Mind's inherent intelligence? Or do we think it's a "jackpot" event that emerged from pure randomness?

Physicalists say it's just chance. Lanza and Berman summarize what the physicalist perspective is asking us to believe:

> The entire universe, exquisitely tailored for our existence, popped into existence out of absolute nothingness. Who in their right mind would accept such a thing? Has anyone offered any credible suggestion for how, some 14 billion years ago, we suddenly got a hundred trillion times more than a trillion trillion trillion tons of matter from—zilch? Has anyone explained how dumb

carbon, hydrogen, and oxygen molecules could have, by combining accidentally, become sentient—aware!—then utilized this sentience to acquire a taste for hot dogs and the blues? How any possible natural random process could mix those molecules in a blender for a few billion years so that out would pop woodpeckers and George Clooney? Can anyone conceive of any edges to the cosmos? Infinity? Or how particles still spring out of nothingness? Or conceive of any of the many supposed extra dimensions that must exist everywhere in order for the cosmos to consist fundamentally of interlocking strings and loops? Or explain how ordinary elements can ever rearrange themselves so that they continue to acquire self-awareness and a loathing for macaroni salad? Or, again, how every one of dozens of forces and constants are precisely fine-tuned for the existence of life? Is it not obvious that science only *pretends* to explain the cosmos on its fundamental level?[14] [emphasis in original]

It seems much simpler, to me, to explain the precision of the universe as a manifestation of the intelligent One Mind. While previously I was sympathetic to the randomness argument, I now see it as its own form of religion. It's like a Hail Mary. Sure, it *could* be right. But it seems very, very unlikely.

Darwinian evolution

While we're at it, why not question the mechanics of Darwinian evolution? Don't worry, I'm not arguing that evolution by natural selection doesn't happen. There's plenty of evidence through fossil records, genetic analysis, and laboratory experimentation to suggest that it's real.[15] The question is: Does evolution occur *randomly*? The answer, according to Bernardo Kastrup, is that—mathematically—we really don't know. Here's a summary of his argument (warning: this is a bit challenging).[16]

Evolution is largely driven by "random genetic mutations" that give rise to corresponding changes in an organism's traits. More simply, if you change a person's genes, that person will be different in some way.

Here is the crux of what we'll discuss. The theory of evolution postulates that changes—"mutations"—in genes occur *randomly*. So, our biologists tell us, *randomness drives evolution*. What we'll see is that no one has any clue—mathematically speaking—whether evolution is actually random.

If a trait is beneficial to an organism's survival, it is likely to be passed on to offspring through reproduction. For example, if a genetic pattern allows our hands to sense heat, that's helpful. If you couldn't sense heat, you might end up injuring yourself and dying. Therefore, the genes associated with "heat sensing" would often be passed on to the next generation because people with those genes would be more likely to survive and reproduce. Those genes would be "selected for." And hence the trait ends up in the next generation and is likely to make it to future generations.

Mutations are critical in this evolutionary process because they help drive trait variation. Variation is what gives rise to so much diversity in biology and even within the human species. According to Darwinian theory, variation arises through the randomness of the mutation process.

"Randomness" is a mathematical concept. Formal tests in information theory can examine a data set and determine whether there are patterns. **But in order to determine whether a phenomenon is random, we need to have a complete data set.**

Bernardo Kastrup offered an example. Assume we have the following number sequence that represents some complete data set and shows a pattern. The numbers go up from 1 to 4, then back down to 1, then back up:

1-2-3-4-3-2-1-2-3-4-3-2-1

Suppose we lost some of the numbers, represented by "[lost]":

[lost]-2-3-[lost]-3-[lost]-1-[lost]-3-4-[lost]-[lost]-1

Then, we'd only be able to see the following data:

2-3-3-1-3-4-1

The set is now mathematically random. But it's only random because we lost numbers from the original set. We can't see the pattern that was originally there, and therefore we cannot judge whether the *original* set was random.

The problem for the "random mutation" theory is that the fossil record has preserved a small amount of the genetic mutations. Our data set is sparse. Furthermore, *we do not have the full, original data set*. We cannot know, based on the data currently available, whether genetic mutations are mathematically random.

Biologists, looking at incomplete data sets, are basically saying, "Look, the pattern is random!" Yes, that's true. But that's not what we're after. We're after whether the original, complete data set is random, not the incomplete one. So the belief in random genetic mutations is, ironically, faith based. This is a *major* issue for biology that needs to be addressed.

Synchronicity

Given everything we've discussed so far, I think we need to question whether *anything* is random. Mathematically speaking, in order to determine whether an event is truly random, we'd need to have the highest bird's-eye view of all reality to see the complete data set. We'd have to zoom out all the way to the Absolute. Otherwise, we might think something is random just because we're looking at an incomplete data set from the Relative view. Unfortunately for us, we don't have that bird's-eye view, so we can't honestly gauge whether anything is random. And even if we did have that bird's-eye view, we'd need a mind with enough computational power to discern patterns.

If the One Mind is as intelligent as postulated so far, isn't it reasonable to wonder if, at some level, *nothing* is random? This would imply that there is meaningful order in all of life to some degree.

Sometimes we're jolted by instances that really make us think about this phenomenon. Psychologist Carl Jung coined the term *synchronicity* to refer to coincidences that just don't seem like coincidences. We often try to brush them off as chance occurrences that are bound to happen sometimes. Do they only occur because we start looking for them? I've asked myself these questions. In my personal experience, the coincidences are sometimes so improbable that I've had no choice but to question my skepticism.

In late 2016, when I began exploring this material, I began experiencing crazy synchronicities all the time. They happened so frequently that I was noting each one in my phone as it occurred. When I analyzed the enormous list, I just couldn't rationally make the mathematical case that all of it was random.

I'll relate several personal anecdotes. In the fall of 2016, one of the first books I read was *The Sacred Promise* by Harvard-trained psychology professor Dr. Gary Schwartz. In the book, he talked about his own wild synchronicities and why they were so improbable that they seemed nonrandom. He also mentioned the Canyon Ranch spa in Tuscon, Arizona. Not only were two of my family members at Canyon Ranch at the time, but a friend from San Francisco was on her way there the next day, and I happened to be wearing a Canyon Ranch T-shirt, which had been given to me as a gift years before (I've never visited Canyon Ranch).

Because of the Canyon Ranch connection, I texted my friend and mentioned the book and the topic of synchronicity. Shocked, she responded with the image of a flyer for a lecture she was about to attend at Canyon Ranch: "Synchronicities & Spiritual Wellness, by Gary Schwartz." I had no idea when I texted her that this lecture existed, let alone that my friend was planning to attend it! We had never spoken about Schwartz before, either.

More recently, I was in a local store and ran into a college-tennis teammate whom I hadn't seen in years. He was a senior when I was a freshman, and I didn't know him as well as some of my other teammates. He and his wife live on the East Coast and were spending a few weeks in the Bay Area for work and happened to be staying in a building on my block. The fact that I ran into him and that he was staying nearby was strange enough. It was a last-minute decision to leave my apartment and walk into that store. I wasn't even looking for a specific item to buy. My friend and I began to catch up, and I mentioned that I had recently released a podcast. When I said the name was *Where Is My Mind?*, he and his wife looked at me, dumbfounded, and responded, "That's the name of the song we walked down the aisle to at our wedding!"

Even the way my podcast came about was synchronistic. In 2017, after I'd written the manuscript for *An End to Upside Down Thinking*, I had the idea to create a podcast on the same topic. At the time, I knew little about podcasts and didn't know people in the industry. The one person I *did* know in the media industry was an old friend from high school whom I hadn't spoken with in several years. He lived in New York and produced sports shows for television. I reached out to him, asking if he had any thoughts on podcasts. I was shocked when he told me that he was about to leave his TV role to join one of the largest podcast production companies. Because of him, I was able to produce an extremely high-quality show—well beyond what I could have ever envisioned.

This happens so often that I'm almost unfazed by it now. But early on, these instances were jarring. My hypothesis is that synchronicities are the instances of nonrandomness that we *are* able to clearly see. All other seemingly random events in daily life are only interpreted as random because of our Relative perspective. If we zoomed out to the Absolute—assuming we possessed the immense mathematical horsepower necessary to discern complex patterns—the randomness would disappear.

Perhaps we tap into synchronicity the more we open up to the One Mind: our filter begins to unlock to such an extent that we are more in sync with reality. Whereas when we are out of touch with reality, we're somehow less connected to the broader stream and less prone to having mystical experiences.

The One Mind Has Free Will

I'm often asked for my opinion on whether we have free will in this nonrandom universe. Before I answer, I want to acknowledge that the questions we're asking are coming from 2D Land, when the full reality is analogously on another dimensional level. So the questions we're asking might not even make sense to begin with.

But if we assume the question does make some sense, here's my answer—and it's not simple.

We first have to define *who* the question is referring to. For example, what if someone asks, "Do I have free will?" I'd respond by asking, "Which 'I' are you talking about?" At the Absolute level (our ultimate identity as the One Mind), it must be the case that it has free will. It is the infinite basis of all reality.

If "I" refers to the Relative whirlpool, however, things get dicey. From the perspective of the Absolute, that "I" doesn't exist as a separate entity. How could something nonexistent have free will? But the Relative does exist in our everyday perceptions (there's a "me" here and a "you" there). And we are connected to the One Mind; we are whirlpools *in* the stream of water. The question then becomes: How does the One Mind's free will relate to and interact with the individual's free will (or lack thereof)?

Let's think about our own lives for a minute. We are born into a certain family, in a certain geography, at a certain time. There are factors in place that we can't control once we're conscious of being in a body. If you unexpectedly get struck by lightning walking to work, could you really control that? The bottom line is: there are aspects of life we can't control as an individual. So in the best-case scenario, we have limited free will at the Relative level while having infinite free will at the Absolute level.

In our daily living, it certainly feels like we have the free will to open a door, to raise our arm, to go to the gym or not, to call a friend or not, and so on. But then we have to ask ourselves where the thought initiating those actions comes from. Try it yourself. Raise your arm. Where did the impulse to do that come from?

Consciousness doesn't come from the brain. What is it that tells me to raise my arm? Where does that thought come from?

It feels like thoughts come from "me" as the individual. But when we examine this, thoughts sort of "appear." It's like an energy arises out of which a thought comes into our minds, and then *afterward* we register, "Oh, there's a thought."

In ordinary states of consciousness, there's a feeling that we control our thoughts and that they come from "me" the person. But when some people have spiritually transformative experiences, even that sense goes away. People in that state identify with the One Mind. While they experience everything the body is doing, the body operates autonomously. It's as if consciousness is activating the body and moving it on its own, with no "individual" doing anything. In Hindu Vedanta philosophy, this is known as "non-doership," the notion that there is no individual "doing" anything.

Here's an example from David R. Hawkins, MD, PhD, who offered a vivid description of his experience in this state (these elevated states of consciousness will be discussed further in chapter 5):

> All things were connected. There was only one life expressing itself with one Self [i.e., the One Mind] through all living things. There was no identification with the body and no interest in it. It was no longer any more interesting than any other body in the room.... It was obvious that the real Self was invisible—without beginning, without end—and that there had been only a transitory identification with the body and the story that went with the identification as an individual.

He then referred to his body in the third person because he no longer felt identified with it:

> The body drove the car to a meeting.... The body spoke to others, spontaneously carrying on normal conversations and behaving in its usual way.... [The body]

seemed to know what to do and did it very effectively and effortlessly.... It seemed like a strange vanity to have once believed that there had been a small self as the author of the body's actions. In reality, the body was at the effect of the universe, and there had never been a doer of its actions....

A thought arose. Now that the way to Reality had been revealed, there could be a return to the consciousness of being an individual person, which had formerly been accepted as real.... The desire to experience the individual self became re-energized on its own. The option of letting it go was present, but there was the return of memory of things yet to be finished in the world. As the sense of "I"-ness returned, the choices were witnessed, not actively decided.[17]

This throws a real wrench into our conventional thinking about "control" over "our" thoughts and actions. Who or what is operating the ship of our individual selves? When people achieve these elevated states of consciousness—when they zoom out closer to the Absolute and unify with the One Mind—there doesn't seem to be an individual with free will. The will only comes from the One Mind, which is the ultimate identity anyway.

My hypothesis is if we mentally align ourselves with the One Mind, the One Mind works through us cleanly. The first step is to acknowledge its existence and our part as a whirlpool in the stream. Most of our society hasn't gotten there yet. Perhaps the more aligned we are with it, the more easily the will flows through us as its vehicle. And maybe that's why highly awakened individuals feel like they aren't "doing" anything, but that doing is happening through them. When we aren't in that state, our individual self somehow obstructs the flow of the One Mind's will. This is just an educated guess from our 2D Land perspective.

How can we align with the One Mind? It can occur through the surrender of our individual will and our aspiration to embody the One Mind's will (more on this in chapter 4). It *feels* like we have the free will to set that intention. But who really knows?

Ultimately, my guess is that at the Relative, individual level, the concept of free will is so complicated that we could never answer it completely from our 2D lens. I certainly don't feel like I understand it. The interaction between the stream and whirlpool simply might not be something we are capable of grasping—as much as we think we can intellectualize it. I'm skeptical of anyone who says he or she knows with absolute certainty how free will works.

And, by the way, all of this assumes that the original question even makes sense in the first place. What we know is that we *feel* like we have control over our actions, thoughts, reactions, emotions, and intentions. We might therefore aspire to think, act, and respond to life as best as we can, with an acknowledgment that we might never understand how decisions or thoughts are being made.

The Golden Rule Is Built into the Structure of Reality

The notion of a "will" moves us in the direction of concepts such as "meaning" and "purpose." It seems natural that life would have meaning and purpose simply given its existence as the intelligent One Mind's manifestation. The exact nature of meaning and purpose might be up for debate and perhaps incomprehensible to the human mind. But, like the 2D Land people in *Flatland*, we can look at data points and at least try to discern the "circle" while remembering that it's only part of the "sphere" (or whatever it is) we can't see.

My take on this is: The purpose of the universe is to *E*volve consciousness. I'm not talking about Darwinian evolution through natural selection (that's why I'm capitalizing the "E"—to differentiate the two). I'm talking about the idea that the One Mind naturally seeks to better itself.

Let's think about our own lives. We are born, we learn lessons while we live, and then we die. If the One Mind model is correct, our individual consciousness continues after physical death.

Consciousness was never tied to the body in the first place, so when the body dies, consciousness transitions into a new state. In this process, we don't take physical stuff with us after physical death. Our clothing, house, and car stay here on Earth. All we take is the way in which our consciousness Evolved. In this sense, the physical universe could almost be likened to an Evolutionary playground for the One Mind to have an infinite diversity of learning experiences through physical vehicles. While veiled from itself, the One Mind's learning is enhanced. It's as if we're living in a fully immersive virtual reality game, where our body is the character we take on.

The life review

The best (indirect) evidence for the notion of Evolution comes from the life review in near-death experiences (as mentioned in chapter 2). As a reminder, near-death experiences have been reported for thousands of years, but the numbers have increased in recent decades due to improvements in resuscitation technology. We can bring people back from the grips of death more regularly; and we get similar (though not identical) reports, across cultures, in adults and in children, across different time periods. They're being studied by top-notch scientists such as Dr. Bruce Greyson at the University of Virginia, cardiologist Dr. Pim van Lommel, University of North Texas professor Dr. Jan Holden, Dr. Raymond Moody, Dr. Penny Sartori, Dr. Kenneth Ring, and many others. They all see essentially the same thing: the life review emerges as a significant pattern (for more, see the "Near-Death Experiences" and "Life Review" episodes of the *Where Is My Mind?* podcast). While not all near-death-experience reports include life reviews, they are reported often enough to warrant serious consideration.

In a life review, people realize that no event or thought in their lives was missed. Every detail was captured. The One Mind misses nothing. Every event is registered since everything occurs within the One Mind, by definition. David Hawkins aptly titled one of his most well-known books *The Eye of the I: From Which Nothing Is Hidden* (2001) for this very reason. The ultimate "I"—the One Mind—has an eye everywhere because it *is* everywhere and everything.

A particularly striking life-review case was that of Dannion Brinkley, the author of *Saved by the Light* (1994), who had four near-death experiences. These occurred as a result of being struck by lightning, open-heart surgery (on two separate occasions), and brain surgery. In each near-death experience, he had a life review that he could still remember.

The next two sentences are potentially life-changing. **Brinkley relived his combat days in Vietnam and experienced the deaths of the people he killed, through their eyes. He also felt the resulting pain of the child who would no longer have a father because Brinkley had murdered the father.**[18]

But, like most people who report a life review, Brinkley also felt the joy he gave to others. His life-review experiences changed him completely, and he became a hospice volunteer. Bringing love and joy to others became his focus rather than materialistic aims. *Evolving* to embody those qualities became the driving force behind his earthly actions thereafter.

Kenneth Ring described another memorable case:

> I have a friend who when growing up was kind of a roughneck; he had a hot temper; he was always getting into scrapes. One day he was driving his truck through the suburb in the town where he lived and he almost hit a pedestrian. And he got very aggravated with the pedestrian and he was a very big physical guy—still is— and a fight ensued. He punched this guy out and left him unconscious on the pavement, got back into his truck, and roared off.
>
> 15 years later my friend has a near-death experience... and during the near-death experience he has a life review. In his life review, this particular scene of the fight takes place again.... And he said that, as many people do, he experienced this from a dual aspect. There was a part of him that was almost as if he were high up in a building

looking out a window and seeing the fight below. But at the same time he was observing the fight like a spectator, he saw himself in the fight. Except this time, he found himself in the role of the other person. And he felt all 32 blows that he had rained on this person 15 years ago now being inflicted upon himself. He felt his teeth cracking. He felt the blood in his teeth. He felt everything that this other person must have felt at that particular time. There was a complete role reversal; an empathic life review experience.[19]

Can you imagine what a mass murderer or torturer encounters during a life review?

Another one of the key emergent themes from the life review is: **The little things are the big things.** Even seemingly insignificant interactions become focal points during the life review. For example, one person who had a life review saw how an unpleasant interaction with a grocery-store clerk put the clerk in a bad mood and impacted how the clerk treated every other person in line afterward.[20]

People come away from the life review with a different perspective on meaning and purpose. The life review is almost a way for us to "grade the test" of our lives and see how we did. Of course we can't know for sure, but we have to wonder if the life review is simply part of at least some people's dying processes.

As former Harvard neurosurgeon Dr. Eben Alexander said about the life review: "It teaches us not to do things to other people that we would not like done to ourselves. **It's basically the Golden Rule written into the very fabric of the universe**"[21] [emphasis added]. At least part of the purpose of life seems to be to embody the Golden Rule, which is an embodiment of our oneness at the Absolute level of reality.

Is it any coincidence that virtually all of the world's cultures and spiritual traditions preach the Golden Rule? It would be easy to explain this if all religions tapped into the same universal reality of the One Mind independently. This would explain why they've drawn the same conclusions. Here are some examples.

Religion	Statement Suggestive of the Golden Rule
Hinduism	"Everything you should do you will find in this: Do nothing to others that would hurt you if it were done to you." (Mahabarata)
Buddhism	"Do not offend others, as you would not want to be offended." (Udanavarga)
Taoism	"The success of your neighbor and their losses Will be to you as if they were your own." (T'ai-shang Kan-ying P'ien)
Native American	"Do not wrong or hate your neighbor. For it is not he who you wrong, but yourself." (Pima proverb)
Confucianism	"Is there any rule that one should follow all of one's life? Yes! The rule of gentle goodness: That which we do not wish to be done to us, we do not do to others." (Analectus)
Judaism	"That which you do not wish for yourself you shall not wish for your neighbor. This is the whole law; the rest is only commentary." (Talmud)
Christianity	"In everything, do to others what you would have them do to you. For this sums up the law and the prophets." (Matthew)
Islam	"None of you shall be true believers unless you wish for your brother the same that you wish for yourself." (Sunnati)

The One Mind Is Made of Love

A corollary here is that the One Mind is not only Evolutionary in nature but is *comprised* of love. Under this idea, love isn't an emotion; rather, it's what the One Mind *is*. It's a natural quality of the One Mind that manifests through vehicles of human bodies. When people experience love, it is a reflection of their tapping into the One Mind—their individuality is melting away, and obstructions to the broader stream are temporarily lifted. This is why in transcendental states of consciousness (such as near-death, psychedelic, and spiritually transformative experiences), a sense of unconditional love, overwhelming peace, bliss, and unity is reported.

The Hindu phrase for this experience is "brahman [the ultimate reality] is satchitananda": "being-consciousness-bliss." When the filter of the brain is unlocked, the underlying love embedded in the universe is felt without restriction. In day-to-day living, we're largely blocked. As Ramana Maharshi put it: "The eternal, blissful, natural State has been smothered by this ignorant life."[22]

The life review is a demonstration that part of the Evolutionary impulse is to fully embody our nature as love.[23] That's why people review the extent of their loving or unloving actions. And we feel this in our daily lives. Most, if not all, of our actions can be tied back to a desire to feel loved or to give love.

We All Have Amnesia

One of the most impactful notions I've encountered is the hypothesis that we already know our true nature as the One Mind made of love—Evolving itself through the machinery of bodies—*but we've simply forgotten this*. Or as Sufi sheikh Llewellyn Vaughn-Lee put it, "We have forgotten that we have forgotten."[24] More specifically, at the Relative level as a human, we've forgotten. But at the Absolute level as the One Mind, we haven't forgotten. We simply don't have easy access to the Absolute level in everyday living while in a body. It's like being in a big mansion, and we are in one room with the light on while the rest of the house is pitch black. The other aspects of the house have always been there; we simply don't have the lights turned on to our True Self.

This is precisely the Evolutionary process we're all a part of: to light up the rest of the house, to re-remember ourselves as an interconnected One Mind after a period of forgetting, and to know our true nature. In other words, we're here *to wake up from the dream*! I've heard it described as the "cosmic game of hide-and-seek."

We have to wonder: Is our amnesia programmed into the structure of life? Perhaps by forgetting, we are able to have a more complete Evolutionary experience in this physical realm. If we remembered everything, we wouldn't be able to learn and grow in the same way. If your teacher gave you all the answers to the test, would you learn? And when we do finally remember after a period of forgetting, it makes the remembering so much sweeter. So we're veiled from our true nature and are slowly waking up to who we've been all along.

Let's examine memory from our own experience. Think about how much you *don't* remember. Do you remember being born? Do you remember being an infant? Or every single thing you did six years ago to this day? Or every single thing you did a week ago? When we think closely about our memory, we realize how limited it is. We actually don't remember most of our lives.

That raises another question: What if there are other important aspects of our existence that we don't remember?

Memory is still a huge mystery in neuroscience. Neuroscientists are trying to understand how memory is stored in the brain and what parts of the brain allow this to happen. I think they're missing the point. They're typically studying memory with the assumption that consciousness comes from the brain.

I don't view memory as something that's stored in the brain at all. The One Mind model suggests instead that memories are in the stream of consciousness, and the brain is a tool to *access the memories*. So certain parts of the brain might be *related* to the memory-access process, but they don't house the memories.

Something about the brain/body system blocks us from parts of the stream. Even in near-death experiences, some people come back to life with distinct memories and talk about a "purpose" that was made clear to them, but they're unable to remember exactly what that purpose was.[25]

In short: The One Mind seems to be veiling itself from itself. And so we all have amnesia and are waking up to the remembrance of our amnesia.

Reincarnation Is the Engine of Evolution

Continuing with the amnesia hypothesis, let's think about the role of reincarnation. As mentioned in chapter 2, there exists strong evidence from more than 2,500 case studies compiled by the University of Virginia (UVA) suggestive of reincarnation (young children with past-life memories and corresponding birthmarks/defects). Even astronomer Carl Sagan, who was typically skeptical of nonmainstream ideas, admitted that this was an area that deserved "serious study."[26]

The work at UVA is strong evidence for reincarnation at the *Relative* level of reality. At the Absolute level of reality, reincarnation couldn't possibly exist. At the Absolute level, there aren't multiple individual beings to reincarnate, and linear time doesn't exist. It's all happening simultaneously. How can there be a "past life" if there is no "past"? The paradox we're forced to hold is that reincarnation both exists and does not. The way we see it depends on the level of reality we're using as a lens.

But if we accept reincarnation to be real at our Relative level, we could view it as the engine behind the Evolution of consciousness. We enter different bodies to have different learning experiences, each time forgetting who we truly are. Sometimes the memories slip through and people remember past lives. Perhaps this is more prevalent in children because they have less conditioning—less getting in the way of the pure stream of consciousness. Since children don't talk in their very early months and years, we don't know what they remember. Perhaps as we age, our developing bodily machinery veils us from things we've always known.

Let's just say we've all had many past lives and simply don't remember them. Wouldn't it be overwhelming to carry the memories in this life? Or we might not have the same learning experience this time around if we remembered. So we forget and have an independent experience. We learn what we can through various challenges and perhaps use the life review to see how we did. Some even believe that when we reach a high level of Evolution, we no longer need to incarnate in the physical world. For this reason, in Buddhism there exists the principle of "getting off the wheel of reincarnation."

Reincarnation implies that there is a nonphysical part of us that continues from life to life. But it's not the One Mind; it still has a sense of individuality. This entity is often referred to as the "soul." So we might consider whirlpools in the stream to be multidimensional, such that we are an individual whirlpool in a body but also having a sense of individuality beyond the body—all within the backdrop of the One Mind.

Skeptic Michael Shermer, among others, contends that in spite of the evidence, reincarnation couldn't possibly be real. One of his arguments is that the population math doesn't work. Throughout human history, billions more people have died than are currently living. So, he essentially asks: "Where are the missing souls?"[27] Another way of framing his argument is: "I know how reincarnation would operate if it existed, and the evidence doesn't match that; therefore, reincarnation can't be real."

A true skeptic would ask: Do we really know how the reincarnation system would work if it existed? And if we accept the One Mind model and its implied immense level of intelligence, do we think reincarnation would be as simple as Shermer is making it out to be?

What if a soul reincarnates many, many times into different bodies? What if souls can reincarnate into more than just the human species? What if reincarnation is broad enough to incorporate life forms on other planets in this massive universe, or even beings in other dimensions of reality? What if the soul inhabiting a person's body is a combination of aspects of many souls combined into one? What if one soul can be split into multiple bodies at the same time? What if some souls can merge with the One Mind and lose their individuality? What if new souls can be created?

The "unknown unknowns" here are endless. So the simplistic "population" argument should be dismissed.

Reincarnation Is Structured

If we accept that reincarnation does exist (at the Relative level at least), then it seems far-fetched to think it would be an arbitrary process. In other words, it seems reasonable to infer that it would have an underlying structure. Given the intelligence of the One Mind, the structure is probably so complex that we could never understand it.

Is there a "planning" process whereby a body is chosen for Evolutionary reasons?

If a planning process did occur, that would imply the existence of the soul in between lives. Researchers at UVA have looked into this idea. Out of their database of roughly 2,500 young children who report past-life memories, 1,200 cases had been entered into a computer system for analysis. UVA researchers Poonam Sharma and Jim Tucker, MD, summarized the results in a paper published in *The Journal of Near Death Studies* (2004): "In 276 of these [1,200 cases], the [child] has claimed to remember not only a past life, but also a time between lives, an intermission memory."[28] Intermission memories included "visions of other beings and suggestions of being in another realm"—both of which are also characteristic of near-death experiences.

The challenge with intermission memories is that they are generally unverifiable. How can we prove that a child saw a "being" in another dimension if we can't see it ourselves? On the other hand, when a child speaks of a previous life, at least the researchers can try to verify the information by finding matching historical records (which UVA often does).

As a proxy, Sharma and Tucker examined whether children who reported intermission memories had more verified past-life memories than cases of children with past-life memories who didn't have intermission memories. That's exactly what they found. While this doesn't prove the existence of intermission memories, it certainly lends more credibility to their claims.

Additional research on this topic has been done by anthropologist

Dr. James Matlock. Most of the cases he examined are memories from young children, and he broke down the reported memories into the following categories:

> The intermission may be broken down into five stages: (1) Death and its immediate aftermath; (2) Discarnate existence; (3) Choice of new parents; (4) Life in the womb; (5) Birth and its immediate aftermath. Memories of all five stages have been reported, although no subjects remember all of them, and most memories relate to Stages 1-3.[29]

All of this is saying: memories before birth exist, and there is at least some degree of planning and choosing before a soul enters a body. From the perspective of everything we've been discussing, this makes a lot of sense.

Similarly, researcher P.M.H. Atwater examined cases of individuals who had near-death experiences as young children—as early as in the womb and as late as five years old. In her book *The Forever Angels: Near-Death Experiences in Childhood and Their Lasting Impact* (2019), she summarized the remarkable findings: "*All of them remember living before their birth*"[30] [emphasis in original].

Some adults recall between-lives periods as well. Many of the cases I've seen come from reports of individuals under hypnosis (known as "past-life regression hypnosis"). If the hypnotherapist is competent, he or she will not ask leading questions, and yet the client will sometimes spontaneously recall past-life memories.

Many reincarnation researchers express doubts about this approach. But even UVA's Dr. Ian Stevenson, who believed in reincarnation based on his research studying children with past-life memories, acknowledged:

> Although I am skeptical about the results of most experiments with hypnotic regression to previous lives, I do not reject all of them as worthless. In a few instances the subject has communicated obscure information about a particular place in an earlier period of history,

> which it seems most unlikely he could have learned normally.... Moreover, in two cases that I have investigated, hypnotized and regressed subjects proved able to speak foreign languages they had not learned normally. This ability is called xenoglossy.... They spoke these languages responsively, that is, they engaged in a sensible conversational exchange with other persons speaking the same language.[31]

The takeaway here is: we should be cautious about drawing strong conclusions from hypnosis evidence, but we shouldn't discount all of it. There are at least some hugely significant data points. If it were completely unreliable, then we would have zero remarkable cases. And yet there are at least some remarkable cases. Something is going on.

We can understand why hypnosis might have this effect. When the client's mind is in a calm state, similar to what happens during meditation, perhaps there is less clutter blocking pure consciousness from flowing in. The filter is unlocked, and more of the stream is accessible. I'd hypothesize that this is another example of "less brain, more consciousness" when done properly.

Between-lives memories emerge often in reports from past-life hypnotherapists and include a degree of planning the current life. Two of the most well-known practitioners in this space are Yale-trained psychiatrist Dr. Brian Weiss (author of *Many Lives, Many Masters* [1988]) and Dr. Michael Newton (author of *Journey of Souls* [1994]).

Hypnotherapist Robert Schwartz's book *Your Soul's Plan* (2009) examined a particularly thought-provoking aspect of prebirth planning: the theory that we plan our biggest life challenges before birth. The implication is that we are never victims in the way we sometimes think we are.

In order to test this idea, Schwartz enlisted mediums who claimed to be able to access the soul of a living individual who had a traumatic life (for example, physical illness, parenting handicapped children, deafness and blindness, drug addiction and alcoholism, death of a loved one, and accidents). The mediums revealed, across multiple individuals with severe life challenges, that many such challenges are planned before birth. Schwartz summarized the findings of his research:

> Ultimately, regardless of the specific challenges they contained, every life blueprint I examined was based on love. Each soul was motivated by a desire to give and receive love freely and unconditionally.... Many souls were also motivated by a desire to remember self-love. Literally, we *are* love.... Life challenges give us the opportunity to express and thus know ourselves more deeply as love in all its many facets: empathy, forgiveness, patience, nonjudgment, courage, balance, acceptance, and trust.... Love is the primary theme of pre-birth planning.[32] [emphasis in original]

While Schwartz's approach certainly was not scientific in the traditional sense, the conclusions of his research do align with themes we've been discussing about the One Mind. By encountering challenging circumstances, we Evolve—even if it's through suffering. In this sense, everyone and everything is teaching us about self-love, which is the quality of our True Self as the One Mind. Self-love is then part of the path back to what we've always been. The people around us—whether we like them or despise them—are actually helping us learn, grow, and Evolve to our ultimate identity as the One Mind. As Ram Dass put it: **"We're all just walking each other home."**

A related account was reported during a *Buddha at the Gas Pump* episode when author Noah Elkrief gave a fascinating description of memories of his prebirth planning. The memories came to him spontaneously when he was an adult and were not obtained through hypnosis:

> I literally watched myself choose and saw how I chose and everything.... All of the sudden I'm watching this screen, and it's kind of like... an iPhone or an iPad where

if you zoom out you can see multiple screens. I could see millions of screens all at once somehow. And I'm seeing all the potential parents that I could have. And somehow when I'm looking at them all at once in no time at all, I know what their themes are. I know what their wounds are, what their beauty is. I know everything without someone telling it to me, I just see it and know it all at the same time, and then I chose my parents from this. And it was so beautiful to see yourself choose.... I'm not a victim.... All the worst things I've ever gone through, I chose.[33]

Finally, something along these lines has been reported under the influence of psychedelics (that is, another instance in which the filter to the broader reality is unlocked). A participant in Rick Strassman's study on the psychedelic DMT (N,N-Dimethyl-tryptamine) reported, "I realized I was in the area where souls await rebirth, and I was there and I had been there so many times before, and this incredible transcendent peace came over me."[34]

Karma Is Real

Given the apparent complexity of the reincarnation system, it feels like a reasonable additional leap to infer that "karma" could play a role as well. Karma is the idea that actions are somehow balanced in the universe. For example, if you harmed someone in this life, maybe in the next life he'd harm you (incarnated in different bodies, without memory of the previous life's interactions).

The notion of a life review makes karma seem like a possibility, and based on all of my research, I do think it's likely that karma is built into the structure of reality. In the life review, we see our mistakes, so it seems to reason that those mistakes would be registered in the One Mind. Perhaps they fuel the reincarnation engine and help lay out optimal future lives for one's consciousness to Evolve.

If karma is real, we should be sure not to oversimplify it. It could be so complex and multifactorial that we'll never understand it. That said, if it is real, it could dramatically change how we think about life.

David Hawkins, like many spiritual teachers, viewed karma as a reality. I mention him here because his description has stuck with me. During his mystical experience of unity mentioned earlier, he said, "The body [that is, his body] seemed like **a karmic wind-up toy** run by all of its accustomed patterns and programs"[35] [emphasis added].

Furthermore, karma is often described as an impersonal process. Hawkins used the analogy of iron filings being attracted to a magnet. If each of us is an iron filing, the way we "charge" ourselves through who we've been and who we strive to become will automatically attract a certain type of magnet (life). So what enters our lives occurs naturally as a consequence of many karmic factors, and not because of some intelligent being who's judging us.

Karma might also create dynamics that encourage repeated incarnations with certain souls. I've often encountered the notion of "soul families": groups of souls who allegedly spend lifetime after lifetime together but take on different roles. And each time, they have amnesia and forget about their previous lives together. Someone's lover in one life might be a parent in the next life, a daughter in the one after that, and then a lover again in a future life. Theoretically, this could explain strong attractions or aversions we have toward certain people that we can't rationally understand. Because we have amnesia, we generally don't remember any of this (although in UVA's studies, some children do report attractions and aversions related to alleged past lives). Sometimes we have a "sense" that we can't put our finger on.

When karma is discussed, a question often arises around tragedy. If someone is murdered, does that occur because he or she did something bad in this life or a past life?

We need to strip ourselves of all preconceptions before examining this issue. First of all, it's impossible for us to judge something as objectively "good" or "bad" because we don't have enough information (more on this in the next chapter). A seeming tragedy could lead to a soul's Evolution and balancing of karma in ways we simply cannot see or compute. In such cases, the "tragedy" has a benefit at some level.

Darkness provides contrast so that we can see light. As twentieth-century American poet Wallace Stevens wrote: "Death is the mother of beauty." When we lack something, we then appreciate its presence more. Perhaps the intertwining of light and dark is behind the yin-yang symbol in ancient Chinese philosophy (below). If the universe had only light, how would we learn and grow?

The Yin-Yang symbol is illustrative of life's intermixed "light" and "dark" elements.

Some object to these notions because suffering can be so great. How could so much suffering be allowed in the world if the One Mind is made of unconditional love? Author David Lorimer stated it well: "We do not have a profound enough philosophy of suffering, especially if we consider that the object of life is happiness rather than growth."[36] I have a feeling that "suffering" is another example of people in 2D Land who only see a circle when in fact it's a sphere.

But does that mean we should ignore suffering, throw our hands up in the air, and say, "I'm not going to stop those murderers from killing people because it's those people's karma to die"?

That is a dangerous attitude, and I strongly disagree with it. To claim we know an event is caused by karma is an immensely arrogant position. Furthermore, author Andrew Harvey suggested in a

Buddha at the Gas Pump episode that while it could truly be some-one's karma to suffer, **if we are aware of that suffering, maybe it's our karma to do something about it.**[37]

CHAPTER 4

APPROACHES TO RIGHT-SIDE-UP LIVING

Approaches and Personal Influences

Now that we've built a more comprehensive picture of reality, we can begin to examine approaches for how we might live our lives accordingly. The approaches described in this chapter logically follow the view of reality outlined so far. This is our next step toward building a compass for living.

In my own life, the approaches we'll discuss have worked. I aspire to embody them, although I'm a work in progress, and challenges inevitably arise (all the time). **Your journey is your own, and you're free to take what resonates *with you* and leave the rest. I'm simply describing here what has worked well for me, while staying in alignment with my view of reality.**

The approaches are meant for everyday living, even regarding events that might seem mundane. That's where the rubber meets the road—in practical applications. The challenge is whether we can execute when life puts us in difficult situations. We can read and theorize all we want, but that isn't the same as living it. Living it can take time and practice, and it inevitably involves trial and error (I can attest to this!).

It's important to keep in mind that the approaches are not black and white. They are subtle, ambiguous, and context dependent. The Buddhist notion of the "middle way" is instructive again: Taking too firm a position could lead to misinterpretations of what I really mean. So, where possible, I try to anticipate potential misunderstandings and clarify in order to avoid confusion.

Influences

During my journey I've benefited from the works of great teachers who have thought deeply about this very topic and given advice based on their own lived experiences. For example, as I mentioned, I've been influenced by David Hawkins (1927–2012); as well as Rupert Spira (1960–present), Francis Lucille (1944–present), Adyashanti (1962–present), Jean Klein (1912–1998), Ramana Maharshi (1879–1950), Nisargadatta Maharaj (1897–1981), Swami Dayananda Saraswati (1930–2015), Swami Sarvapriyananda (birth date unknown, but he is still living), Joel Goldsmith (1892–1964), Huang Po (died 850), Eckhart Tolle (1948–present), and others.

Additionally, the *Buddha at the Gas Pump* podcast—hosted by Rick Archer and including more than five hundred free episodes of conversations with ordinary spiritually awakening people—has had a major influence. I've listened to hundreds of hours of that show and highly recommend it to anyone on a path of personal and spiritual growth.

By using this multipronged approach, I've been able to go deep with specific teachings while simultaneously maintaining significant breadth. What I've found is that people at advanced levels of consciousness—across cultures, traditions, and time—often describe a similar view not only of reality, but also of practical living. *There is an undercurrent of universality.* People might use different language, but they're basically talking about the same thing. What you'll see here is my way of synthesizing and describing these ideas, but they really shouldn't be considered "mine."

This section heavily features the work of psychiatrist-turned-spiritual teacher David Hawkins. He coauthored *Orthomolecular*

Psychiatry (1973) with Nobel Prize winner in chemistry Linus Pauling, and ran a leading New York psychiatric practice. As a teenager, Hawkins claims to have spontaneously experienced the totality of human suffering and wondered what kind of God could ever allow it. He then immediately became a devout atheist into his thirties before profound mystical experiences shifted his life completely. Because of his background in psychiatry, he saw what worked for people at a very practical level and specialized in deconstructing the human ego. For that reason, I've been particularly drawn to his psychological perspectives.

With that preface, below is a list of ten approaches to life that we'll review *in light of the picture of reality discussed so far*:

1. Nonjudgmentalism
2. Surrender
3. Nonresistance
4. Nonattachment
5. Forgiveness
6. Compassion with discernment
7. Authenticity
8. Stewardship
9. Nonconceitedness
10. Commitment

Nonjudgmentalism

Let's reestablish a few ideas. First, since we know almost nothing, we need to be radically humble about our assumptions about *everything*. Second, evidence suggests that we are part of an unimaginably intelligent One Mind, made of unconditional love, which has some sort of will and underlying Evolutionary drive. We might not know the mechanics of all this, but we acknowledge that there's potentially big stuff happening beyond what our intellect realizes. Therefore, we can't know what the One Mind "wants" (assuming it even "wants" something in the first place).

In our society, we tend to evaluate events and people as "good" or "bad." I ask you: Who are we to judge? What do we know?

Let's take an example of something most people would consider "bad": the death of a loved one. In a hypothetical situation, Jill's husband suddenly dies of a rare illness, leaving her alone with three kids. Jill is understandably devastated, depressed, and overwhelmed. But two years later, she unexpectedly meets another man, falls in love, and remarries. Her new husband is very wealthy and agrees to fund research to cure the illness that killed Jill's first husband. Eventually, due to this funding, a cure is found, saving many other lives. Inspired by all of this, Jill's children become doctors and devote their lives to preventing illness in others.

So, was the death of Jill's husband "good" or "bad"? Yes, he died, resulting in temporary suffering for his close family, but in the end, many people were helped because of his death, and love was spread.

Judging this event depends on the lens from which we view it. It's simply impossible for us to see the ripple effects of any individual event from the Relative perspective. What seems like a tragedy might be assisting with someone's Evolution in ways we can't understand. And who knows what past-life karma might be involved or how it relates to the One Mind's will?

I think of it as if we're running through a maze, only able to see what's immediately in front of us. If we had a helicopter's view that could see from above and outside of time, like the One Mind does, we could see the full maze. We would see what's ahead in the maze and how an event fits into the bigger picture. But since we don't have that view in the Relative, we simply have to let go of our instinct to judge everything.

This principle, known as *nonjudgmentalism* (a term often used by David Hawkins), does not imply that we shouldn't have preferences or desires. Since the One Mind is made of unconditional love, we might assume that anything leading to unconditional love is somehow "preferred" by the One Mind.

We simply need to acknowledge that labeling things as "good" or "bad" is actually a form of arrogance. It's like saying, "From the vantage point of my little whirlpool, I know what's best for the stream." We have no idea what the One Mind is actually up to, so it's best to drop overly rigid positions. That's not to say we should become emotionless robots. I'm not arguing for any less engagement in the world. I'm just noting that humility implies a loosening of the grip on our tightly held personal opinions.

Surrender

A related notion is that of surrender, which is the attitude of acknowledging that while we are aspects of the One Mind in the Absolute... in the Relative, we are puny whirlpools compared to the full stream. There are forces at play beyond all comprehension, and therefore the best we can do is set the goal (an intention) to be in alignment with the One Mind—whatever it's up to—and surrender to that intelligence as being far superior to ours. As Adyashanti says, at a certain point our ego realizes just how limited it is—then it realizes it is "checkmated."

It's as if we're floating in a river and we can either try to impose our personal will to fight the current—which is ultimately a losing strategy, leading to lots of flailing and suffering—or we can attempt to align with the current and let *it* carry us. This is where the religious saying "Surrender your personal will to Divine Will" comes from. It's out of humility that we surrender. We can't know if what the individual is trying to impose is what's "optimal" from the Absolute perspective.

This attitude is one of *allowing* rather than *forcing*. There is, of course, a fine line here. Surrendering doesn't mean being lazy and saying, "I'm not getting off my couch because I'm surrendering to the One Mind." Surrendering isn't a justification for irresponsible behavior or dropping obligations and duties. A balance is required, echoing Paramahansa Yogananda's notion of being "calmly active and actively calm." I'm advocating for an active passivity and passive activeness.

So, how do we know when to *act* if we're "surrendering"? I wish I had a black-and-white answer, but I do not. If it were that simple, everyone would be doing it. But here are four guiding principles I use for acting in the world while maintaining an attitude of surrender. This is admittedly a process of trial and error, which can only be fine-tuned by testing it out repeatedly in your own experience.

1. Follow your values (for example, integrity).
2. Follow what feels viscerally passionate—a no-doubt *yes* (it's a feeling that only you know for yourself).
3. Follow your intuition (an indescribable sense of knowing that goes beyond reason).
4. Do "the next obvious thing" (a phrase adapted from spiritual teacher Suzanne Segal[1]).

We might even view our passions and intuition as signals from the One Mind that we're on the right track. And when a situation presents itself with an obvious next step, that itself could be silent guidance from the intelligence of the One Mind. For example, I had been loosely thinking about ideas for a second book for a while but never felt an impulse to embark on this task. I simply kept researching the topics that interested me, but I didn't start writing. And then one day, the instinct came to begin, at which point I furiously starting writing with 100 percent passion. I worked on this book nonstop for a week and completed the first draft of the manuscript. Through an attitude of surrender, I "allowed" it to come to me and didn't write for a while. But as soon as the "hit" came, I acted with full force. I was both passive and active.

Surrender can also be useful when we become anxious about the future. Sometimes we tend to project too far ahead, at which point surrender can kick in and we can relax our thinking, focusing more on what is immediately in front of us.

Thinking back to my tennis-playing days, living in a surrendered state is much like playing a match in a "flow" state. I played my best tennis when I let the match come to me, played loosely, and simply followed my instincts—all while giving 100 percent effort.

On the other hand, I didn't play as well when I felt like I was "trying too hard" or "forcing" the situation.

Furthermore, a surrendered attitude toward knowledge might (paradoxically) have the benefit of allowing us to absorb more of the One Mind's intelligence. Hawkins contended: "One has to see through the mind's illusion that it knows anything. This is called humility and has the value of opening the door for realizations, revelations, and intuitive knowingness."[2]

Nonresistance

When we surrender, we also stop resisting life. We allow emotions to flow, rather than suppressing them. We *let things go*. This is *not* what we are conventionally taught.

David Hawkins described this process in his classic book *Letting Go: The Pathway of Surrender* (2012):

> Letting go involves being aware of a feeling, letting it come up, staying with it, and letting it run its course without wanting to make it different or do anything about it. It means simply to let the feeling be there and focus on letting out the energy behind it. The first step is to allow yourself to have the feeling without resisting it, venting it, fearing it, condemning it, or moralizing about it. It means to drop judgment and to see that it is *just* a feeling. The technique is to be with the feeling and surrender all efforts to modify it in any way. Let go of wanting to resist the feeling. *It is resistance that keeps the feelings going*. When you give up resisting or trying to modify the feeling, it will shift to the next feeling and be accompanied by a lighter sensation. A feeling that is not resisted will disappear as the energy behind it dissipates.
>
> As you begin this process, you will notice that you have fear and guilt over having feelings; there will be resistance to feelings in general. To let feelings come up, it is easier to let go of the reaction to having the feelings

in the first place. A fear of fear itself is a prime example
of this. Let go of the fear or guilt that you have about
the feeling first, and then get into the feeling itself.[3]
[emphasis in original]

Hawkins often said that if you're feeling something, you aren't
feeling enough of it. For example, if you're angry, that's because
you're resisting the anger that wants to come out. Let out all
the anger you have, and yell and scream as much as you want
(probably best done in private) until the energy behind the anger
dissipates. If you feel sadness, maybe you need to allow tears to
come out (even if you don't know why). If you feel worried, let
all of your symptoms of worry come out... and so on. Many of us
have layers of suppressed emotions that are bubbling beneath the
surface, probably wreaking havoc on our lives in ways we don't
even realize.

The notion of nonresistance applies not only to emotions, but also
to life situations. If we don't like a situation, we have to ask ourselves why. What are we judging? What could we be surrendering
more of? Of course, nonresistance is not the same as nonaction;
it's simply an attitude shift. With nonresistance we can recontextualize anything that happens, *flow* with it, and make the
best of it. We can view *anything and everything* from the lens of:
"What am I learning from this, and how might it help me Evolve
my consciousness?"

Nonattachment

This is a big one and might be the most difficult to implement
practically.

A natural consequence of surrendering and letting go is that we
gradually become free of attachments. "Attachment" is an attitude
of "clinging-ness" and "craving-ness." It is a mental requirement
that an outcome be a certain way. This is distinct from desire,
which is simply wanting something to happen without an obses-
sion with the outcome. Desires are fine. Attachment, on the other
hand, means that you're *bound* to your desires. Fears are the flip

side of desires (the desire that something *not* happen). So non-attachment is the avoidance of being bound to desires and fears.

As David Hawkins and other spiritual teachers often mentioned, *this is not the same as detachment.* Detachment is a total lack of caring, which is its own form of arrogance. For example, to detach from the world and ignore the suffering of fellow humans would be in opposition to the compassionate quality of the One Mind. Watch out for this; it's an easy trap to fall into on the spiritual path.

So again, we're dealing with a fine line. The preferred attitude of nonattachment is one in which we are active in the world, give 100 percent effort toward whatever we are doing—with full passion—but have no attachment to what ultimately happens or when it happens. The quality of patience naturally emerges as a result.

This reminds me of strategizing for tennis matches. All I could do was execute a strategy to the best of my ability. How the opponent handled it or how the wind blew were factors beyond my control.

Since we have limited control over outcomes, to attach to an outcome is to set oneself up for disappointment. Attachment is like saying, "If X doesn't happen, then I'm going to be miserable." Why create that game for yourself?

Furthermore, attaching to an outcome presumes that we know what outcome "should" happen in the context of the One Mind's intelligence, karma, and whatever other factors are at play. The principle of radical humility makes notions of attachment seem absurd.

In my own life, and in many of my friends' lives, attachment is usually a key factor behind any problem. If someone comes to me with a problem, before even starting the conversation, I'm inclined to just start asking, "What's the attachment here?" Nearly every time, the problem is related to being extremely attached to something. From there, we can examine what the attachment is, why it exists, and whether it's warranted. Usually the intellect starts to realize that the attachment is some form of conditioning, which, upon honest inspection, can be identified as irrational and stems from a lack of humility. With willingness, one can relinquish the attachment. It's not always easy, though.

Attachments can be sneaky. Sometimes the attachment is to a sensation that an emotion gives us—even negative ones. Hawkins advised that we let go of and surrender the "juice" that we get from an emotion. For instance, "resentment" or "guilt" can cause negativity, but secretly a part of us enjoys—and "juices"—the payoff from the emotion, almost like an addict. If we want to get rid of negativity, the solution is to first identify the related emotion and then let go of the attachment to the juice we get out of it.

Manifesting

Along these lines, I'm often asked about the concept of *manifesting* desires—a notion popularized by the book *The Secret* (2006). Given that all reality is just consciousness, by shifting our consciousness, wouldn't reality then shift in some way? Wouldn't reality almost be malleable? We see that this is true in psychokinesis studies (chapter 8 of *An End to Upside Down Thinking*): our minds can shift matter, even if the effect that we measure is sometimes small.

The egoic mind then thinks, *How can I use this to get the car or spouse or job I always wanted?* "Manifesting" is the act of getting those desires through the mind's manipulation of reality. There is an increasing number of personal development coaches who help people do this through techniques such as visualization and feeling the outcome before it happens.

I am extremely cautious about this. That's not because I think it doesn't work; I actually think it *can* work sometimes. **The problem is that it assumes that we know what we want.** Psychology suggests that we sometimes *don't* know what we want, as evidenced by the study of "affective forecasting errors" (a term coined by Timothy Wilson and Daniel Gilbert in the 1990s). Here's the basic idea, as summarized in *Psychology Today*:

In Wilson and Gilbert's research, they found that people misjudge what will make them happy and have trouble seeing through the filter of the present. They also discovered that how people feel in the moment blinds them, coloring the decisions they will make down the road. Unfortunately, people cannot accurately take into account how they might feel in the future; instead, they tend to overestimate how positive or negative they would feel about future situations. Another example: When a person wants a certain item, such as a luxury car, that person anticipates immense extended joy. However, over time, that joy of owning that car will dissipate.[4]

We also have to be mindful of the old adage "Be careful what you wish for." Maybe you try to manifest a huge house and succeed. Now you're living in the huge house, but there are unanticipated headaches you now have to deal with, such as maintenance, cleaning, technical problems, and more. Also, since the house is a bit ostentatious, it has attracted burglars. Your friends start treating you differently as well. The list could go on and on. We sometimes forget that there can be unwanted side effects.

But the biggest issue with manifesting is that it is subtly arrogant. It is an imposition of the individual's will, with its limited intelligence and perspective. The One Mind is infinitely more intelligent than the individual (even though, paradoxically, the individual *is* also the One Mind). Why not surrender to that broader intelligence, which knows much more about our Evolutionary journey than we do? On top of that, our ability to dream up a future on our own might be way more limited than what is possible. So by trying to manifest something specific, we potentially shortchange ourselves because of a lack of imagination, confidence, and/or vision.

Furthermore, when we say we want a "thing," what we really mean is we want a "feeling." We view the thing as the portal into that feeling and mistakenly latch on to that thing. Why not go straight to the source and aim for the feeling? Ask yourself: "What is the feeling I'm trying to experience?"

Inevitably, the answer will come back to a desire for love and its associated qualities of peace and bliss. Looking for the perfect spouse? Maybe that's because you want someone to love you, and you want your friends and family to approve of you (that is, you're looking for their love). Then you can be happy and at peace. Looking for the perfect house? Maybe that's because you want a reward for the hard work in your career and you want to impress other people (that is, you want to express love for yourself and feel love from others). Want better health? Maybe that's because when you're healthy and vibrant, people love you more and you can be at peace.

Any desire we have is, at its core, a need for one of the fundamental qualities of the One Mind. That is so because we *are* the One Mind, veiled from itself, seeking itself. We mistakenly seek out objects (activities, relationships, accomplishments, and so on) to fulfill ourselves. But they are the middlemen between where we are and what we actually want.

So, if the One Mind is the highest state of existence, why not try to manifest its qualities? Then we can attempt to manifest the *feeling* and allow its eventual form to be what it will be. Maybe what you end up getting is beyond what you could have ever imagined.

I'm being intentionally harsh here toward traditional manifestation because, if misused, it can be dangerous and lead to entrapments of the ego. If manifestation is used perversely, it is using spirituality to bolster the ego, which is counter to the Evolutionary process.

That said, we're only human. Desires naturally come up, where you say, "I'd really love to have *that*." You simply can't control it. So in the spirit of nonresistance, we *fully* allow the desire to be there. We sit with it, and we don't fight it. Once we evaluate whether the desire is egocentric or something worth pursuing, we can determine whether to follow it. But we don't attach to it.

Prayer

For the same reasons I'm cautious about manifesting, I'm also cautious about "prayer." They are similar in that they both involve creating an "intention" for something to happen. I like Hawkins's advice, which was to "live your life like a prayer." You *become* what you pray for the world. You act it out yourself.

Prayer has more of an "asking" flavor than manifesting, however. Manifesting involves making something happen by altering reality, whereas prayer is a request. And therefore, prayer is a more humble approach: the act of asking acknowledges one's limitations relative to the intelligence of the One Mind.

I often hear in spiritual circles that an attitude of "asking" for help is required to activate help. I distinctly remember hearing this on the *Healing Powers* podcast episodes back in August of 2016. Thinking back, I tried it out myself. At the time, I was experimenting with sensory-deprivation flotation tanks, in which one floats in a tank full of salt water with the lights turned off and no sound. I had heard about this on podcasts and decided to give it a try. Basically, it's like "meditation on steroids." At the time, my worldview hadn't shifted yet. I still had no belief that life had meaning or that the supernatural existed. I was simply exploring.

I now remember something I had almost forgotten: While I was in my state of despair, I did what I heard on *Healing Powers* and asked—out loud while in the sensory deprivation pod—for help. Is my resultant life turnaround a response to that? Who knows. It took very little effort to try, but because of my desperation at the time, I think my request was genuine. And notice that I didn't say, "Help me: I want a new job, a new relationship, improved health, and a new apartment while we're at it."

As we advance, the highest form of prayer becomes not to ask for any*thing* at all, but rather to fully surrender to the One Mind's intelligence and will. Prayer thus transforms into a willingness to serve because we reach a point where we stop craving anything from the world (via nonattachment). With humility, we "ask" the full stream of consciousness how we can best be of service.

Death

The biggest attachment for most of us is the attachment to our own lives. That occurs when we treat our individual body as the sole source of our identity. I'm reminded of the popular acronym "YOLO," which stands for "You only live once." Well, obviously, I think this phrase is incorrect. It's yet another example of how embedded physicalism is in our Western culture.

When we zoom our perspective out, we can see that this life is only a blip in the context of timeless eternity (the Absolute). We begin to acknowledge that even when we die in this life, we still remain infinite beings on a different time horizon. The images that follow are visual representations:

The Mainstream Western View of Life
Finite timeline with a beginning and end

The One Mind View of Life
Infinite timeline

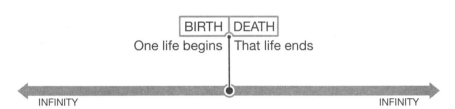

Since we are infinite beings, an individual incarnation is a tiny portion of an infinite timeline. When we understand this, we begin to look at this life in the context of a different time horizon with the knowledge that even when we die in this life, we still remain infinite beings.

Coming to grips with death is one of the most important things we can do. Most, if not all, fears can be boiled down to a fear of death. That stems from the belief that we are finite beings and forgetting that we are actually *infinite beings who never die*. The fear comes from a forgetting of who and what we are.

To illustrate this notion, David Hawkins often encouraged us to play the "And then what?" game. Let's say someone is afraid of losing her job. You'd say, "Okay, you lose your job, and then what?" Then I wouldn't make any money. "And then what?" Then no one would love me, and I'd be homeless. "And then what?" Then I'd starve, get sick, and die. Hawkins would then say, "So?" Laughter usually followed.

This is a tongue-in-cheek method of illustrating the implicit attachment we tend to have to this body and this life. We often avoid the topic of death because we're conditioned to think it's depressing. That's based on a worldview of impermanence and separation. When we fully internalize the idea that we're infinite beings, forever connected, death loses its punch. In fact, many near-death experiencers report completely losing their fear of death and wish they were still in the other dimensions because of how pleasurable they were.

To be very clear: I'm not saying we should be stoic and not care when someone dies or not care about our own lives. That would be to resist our natural feelings, which is the opposite of what I'm arguing (remember nonresistance!). I'm also not suggesting that we suddenly engage in reckless behavior. I'm simply suggesting that we would be well served to lighten up and cling less to this body and this incarnation. Life in this context can be viewed with more humor and equanimity and less unnecessary seriousness. Some call this attitude "the divine indifference." As Saint Francis of Assisi (1181–1226) wisely advised: **"Wear the world like a loose garment."** When we live in this way, we can more fully enjoy life and have more fun.

When we relinquish the fear of death, we can be truly *liberated*. In fact, the letting go of our attachment to life can result in intense

awakening experiences that enable us to live more fully in alignment with the One Mind (more on this topic in chapter 5).

Forgiveness

By forgiving others, we help ourselves. Holding a grudge and
seeking revenge creates an energy that drags us down. Furthermore, we never know why people act in the way they do. We're
not living our lives in their whirlpools. Presumably, they're acting
in ways that seem optimal to them, given their life circumstances,
over which you have no control. What you *can* control is how you
react to them.

With an attitude of humility, we remember that if others seem
to be harming us, we can't know what Evolutionary or karmic
impact that is having on us or them. What might have seemed
"bad" is actually "helpful" in ways we can't see.

And furthermore, from the Absolute perspective of the One
Mind, forgiveness is obvious because the distinction of "you" and
"me" disappears. "You" *are* "me." At the deepest level of reality,
the consciousness that looks through your eyes is the same as the
consciousness that looks through my eyes. As I've often heard it
stated: "There's only one of us here."

Two episodes of the *Buddha at the Gas Pump* podcast stand out to
me on this topic. One featured a woman named Isira, who *became*
the man who was murdering her while he was murdering her (he
did this because she wouldn't accept his advances). Her account
sounded similar to what happens in a life review, except this one
was like a "real-time" version. Isira described the brutal situation:

> He was literally trying to kill me and stating it—that
> he would absolutely kill me. I was not going to get out
> of there alive. My life was on such a thin edge, that I
> knew that if I did not harness every degree of my con-
> scious presence, I would die. So I completely focused.
> Absolute, one-pointed awareness. And in that moment,
> I literally became everything. I became the room, I
> became him, I became the space, I was the body being

murdered, I was everything. And in that I was able to see that man's soul and consciousness. And the reason he was doing what he was doing was because he was so disconnected. He was broken and without love. And in that woundedness he was so desperate for connection and so hurt from not having had it, that he went about it in a way of violation.

That was his own violation that he was acting out. And in seeing that, I just felt the most incredible, immeasurable love for this soul. It was endless, this love, just a wish and a prayer that he could remember and know that he is love, that he is lovable, that he is loved. And literally in that moment the entire room lit up. And it was literally like a thousand angels were in the room. And that man, he looked up and he saw this light. And he dropped to his knees and said, "Oh my God, what have I done? I've killed an angel." He pleaded for my life to be saved. He literally had an awakening. So he picked up my body, called for emergency.... He realized what he had done, so on the one hand he had an awakening... on the other hand he realized what he had done so he [ran away].... I ended up in a coma.[5]

In another astounding *Buddha at the Gas Pump* episode, Julie Chimes described being stabbed to death by a psychotic woman, and during the murder, Chimes had a mystical near-death experience. In the struggle to save herself, Chimes had an opportunity to turn the knife back on the murderer. But she couldn't do it because in this alternate dimension of reality, she felt love and compassion for the murderer. Chimes ultimately survived and concluded the following about forgiveness: "I feel that now there is nothing to forgive.... It's part of a design that was exquisite. For me, for her probably, and everyone involved. It was part of a soul's design. It did have to happen.... I've come to a place of great peace and understanding about that. I really don't think about this woman [anymore]... I don't get anything other than a sense of I just wish her well in her life."[6]

We can perhaps more fully understand Jesus Christ's famous quote: "Forgive them, for they know not what they do." Translation: *When people are harming you, it's because they think of themselves as separate and forget we're all connected as part of the One Mind. If they knew we were part of the same stream, they wouldn't do what they're doing. Their amnesia is thick.*

Compassion with Discernment

Closely accompanying forgiveness is the related notion of compassion for others. If we view ourselves as interconnected and have empathy for the challenges of being human, we can assume a more loving attitude toward everyone. After all, in the life review, what seems to matter is whether we acted from a place of love. Orienting in that direction can help us recontextualize many otherwise difficult situations in life and enhance already positive ones.

We do our best to avoid injuring others and instead aim to help. We aspire to be altruistic. From the perspective of the One Mind, by helping others we are helping our "self." Altruism can thus be viewed as the highest form of selfishness. Furthermore, we literally benefit ourselves by helping others: personal experience and neuroscience teach us very clearly that helping others makes us *feel good.*[7]

I like the simple but profound advice that David Hawkins often gave. This alone is enough of a directive to guide all behavior: **"Choose to be easygoing, benign, forgiving, compassionate, and unconditionally loving—toward all life, in all its expressions, without exception, including oneself."**[8]

Hawkins often used the example of seeing a beetle on its back on the ground. The attitude of compassion compels you to take five seconds and turn the beetle over with a twig rather than step on it. That's a compassionate orientation. The seemingly small act is a reflection of something much bigger: who you *are* in the world. That's really the best any of us can do.

We need to beware here, however. In aspiring to be compassionate, forgiving, and altruistic, we still need to set boundaries and protect ourselves. Each of us individually is part of the One Mind, so to ignore your own well-being is actually a disservice to the One Mind. Trying to pet a barracuda isn't prudent. So, prioritizing spiritual aims doesn't mean being foolish and ignoring self-defense or your own needs. Common sense doesn't go out the window just because you're trying to be spiritual!

Here is an extreme example from a discussion with Pernilla Lillarose on the *Buddha at the Gas Pump* podcast. As she began entering higher states of consciousness, she developed a loving attitude toward everyone. She was in a beautifully surrendered state. In that state, she met a stranger whom she let into her home. He lived with her for months, took advantage of her, manipulated her with religious justifications, and tragically began to run her life. She commented, "It was absolute hell for five months." Lillarose now openly admits: "I couldn't discern."[9]

Being loving and compassionate doesn't mean we have to say yes to everything. If you say yes when you mean no, you not only make yourself uncomfortable, but you give the other party inaccurate feedback about his or her behavior. The whole system suffers as a result of your inability to set boundaries and do what is best for you. Of course there's always a fine line between selfishness and boundary setting, so the type of action taken will depend on the situation's specifics. The boundary should be set in the most caring way possible in order to avoid potential injury.

Anita Moorjani (the author of *Dying to Be Me* [2012]) often talks about this general concept. She was dying of cancer and had a transformative near-death experience. In that altered state, she realized, among other things, that she had been too concerned with what others thought and didn't say no enough. When she shifted her mind-set, her cancer miraculously disappeared, and her doctors were stunned. As she now says, "Love yourself more by saying no to things that feel draining and burdening for you. Saying no when you want to say no, instead of saying yes to please other people, that's a true act of self-love."[10]

Boundary setting can be challenging in a culture where pleasing others is so important. Also, the answer isn't always a clear yes or no. This is yet another process of trial and error, and one I've struggled with. I can't emphasize enough how critical it is, though.

Authenticity

When we're compassionate toward ourselves, we can be more of who we actually are. What else is there to be in this life? Each of us seems to be here for some purpose as "sense organs of the Infinite," so to resist what feels authentic to us might actually be contravening the Evolutionary impulse of why we exist in this body.

Let's use an analogy to dive into this further. Imagine that the One Mind is the sun.[11] The sun's rays are blocked by clouds. In the context of life, "clouds" might be our social conditioning, thoughts, and so on. The more we can get the clouds out of the way, the more rays of pure consciousness—the One Mind and its intelligence—can flow into us. The more rays we let in, the more authentic we are. The more we block the rays, the less authentic we are.

When we have lots of clouds blocking us, we feel like we're faking life. And when we're faking it, we're not living in integrity. And when we aren't living in integrity, we create subconscious guilt. Walking around with guilt is not fun.

In today's society, I feel that inauthenticity is rampant. There-fore, there's an overwhelming amount of subtle guilt and lack of integrity. People often act because of what society or their family tells them they "should" do and not because of a natural passion or intuition. It's difficult to live that way for too long. I wonder if many health problems are caused, at their core, by underlying inauthenticity. The body becomes weakened trying to fight the uphill battle. It takes energy to be something other than who you are! Moorjani's cancer might be an example of that. And through her illness, she learned an incredibly valuable lesson that she's spreading to people all over the world (another example where a seemingly tragic event led to broader good).

Over the last few years, I've had to confront parts of myself that haven't been fully authentic, but recently I was faced with a true test. By 2018, I had advanced to Partner at my firm, which I had worked hard for since joining in 2010. I was in an ideal professional situation with terrific colleagues and fascinating projects and clients. But by late 2019, I started to realize that working at my day job was inauthentic. My passion simply wasn't there like it used to be, and that caused anxiety. I realized that because my true passion lay elsewhere, it was inauthentic to my colleagues and my clients to continue in that position. So I decided to transition out of my role at the firm.

It was a very difficult decision, but it was a fully authentic one. Even though I felt as if I were jumping off a cliff, it didn't feel courageous. It just felt like the right move. As I began transitioning out of that role, my anxious energy dissipated rapidly. In fact, shortly after making the decision, the initial ideas for this book started to enter my mind. It's as if my inauthenticity was blocking my access to unobstructed consciousness; the whirlpool suddenly became more permeable such that more water from the broader stream could flow in.

It felt like a game of Jenga. The tower of blocks represented the inauthentic self I had built throughout my life. Since this journey began in 2016, I was pulling out blocks of inauthenticity, and the tower's stability was becoming tenuous. Transitioning out of my Partner role was like pulling out the block that made the tower fall. But in Jenga, there are usually a few blocks at the bottom that remain even when the tower falls. Analogously, I still have deep-seated aspects of feeling like a separate self rather than as a part of an interconnected unit of consciousness, and probably others as well. The layers—my "masks" of inauthenticity—are slowly subsiding, however.

I mention my personal story with caution. I wouldn't want someone to read this and hastily leave his or her job without careful thought. Given the particulars of my life situation, the move was appropriate. It also came after *more than three years* of exploring my passions on the side. Rick Archer, the host of *Buddha at the*

Gas Pump, has often referenced his personal story on his show: while the podcast has been his true passion for many years, he also worked a day job simultaneously until the podcast became self-sustaining.

But I do want to urge each of us to look hard at all aspects of our lives and think about what feels aligned and what doesn't. What makes us feel truly authentic, the truest version of ourselves? And where are we needlessly faking it?

As I've stepped more into my "real" self, I've noticed that I'm more sensitive to, and intolerant of, inauthenticity than I used to be. It's sort of like when you go on a cleanse and then go back to eating your old food—suddenly your body can't handle it anymore. As spiritual teacher Adyashanti says, when we engage in misaligned activities while on the spiritual path, it starts to feel like "rat poison." We become intolerant of those sorts of behaviors and relationships. It's as if the rat-poison feeling is the One Mind's way of steering us away from something and toward something else.

I've found that relationships shift once one starts to make these changes. Old relationships that used to be in alignment start to fall away. That's completely natural. It's only a problem if we *resist* the feelings of misalignment.

The old notion of "keeping holy company" definitely helps. Stated another way, one's path can be assisted by spending time with people who share similar values and life priorities. Simply put, try to find like-minded friends, and don't sweat it if old relationships change!

Stewardship

When we're fully authentic, we're best able to express our unique talents and abilities. Each of us has them. Think about it: regardless of who we are, there are aspects of our being—no matter how seemingly big or small—that have the potential to make a uniquely positive impact on others in some way. Maybe it's through art, cooking, parenting, financial analysis, medicine, sports, cleaning, clerking—anything. These are our "gifts." Thus, each of us has a way of making a contribution to the world, without judging whether it's a "big" or "small" contribution (remember nonjudgmentalism!). We don't have the Absolute view to discern whether anything is "big" or "small" in the grand scheme.

If we're interconnected, what better use of our skills could there be than to help others in some way? Thus, we have a responsibility to the One Mind—the truest version of our self—to use our skills to their fullest benefit. That could mean reprioritizing one's life. In other words, we should be true stewards of our skills. This is what has happened for me: I'm now using skills developed academically and professionally to help others through writing and other media.

The same goes for money and resources: if we have a lot of them, we should be stewards because they are ways to potentially benefit society if allocated properly.

We're also stewards of our bodies. I now view the body as something I'm borrowing for this life, whereas I used to view it as "me" and "mine." I look at it more like a car I'm renting. This attitude has inspired me to focus on nutrition and healthy habits, such as careful eating and exercise. To ignore proper treatment of the vehicle would be disrespectful, since it is this very machinery that allows me to experience the physical world.

To carry this analogy further, the healthier the vehicle, the purer a vessel it can be for the One Mind, and the better we can express our true being. So health and wellness naturally become top priorities out of respect for the One Mind: We feel gratitude for having this vehicle and aspire to use it for the purpose of serving the Evolution of consciousness on the planet, in any way we can.

On an even broader level, we are stewards of the planet, which enables us to have this Evolutionary experience and houses other life forms that are part of the One Mind. To treat the planet poorly is an act of arrogance.

Nonconceitedness

If we value and express our unique gifts and talents, there can be the risk of excessive pride (conceitedness). It is derived from the belief that "we are so special" and we take undue credit for our accomplishments. When we have the humility to acknowledge how many cosmic factors are likely at play behind the scenes, we loosen our sense of ownership over achievements.

But even if we see things from a nonspiritual perspective, what we accomplish in life is caused by many worldly factors beyond our immediate control. Swami Dayananda Saraswati summarized this idea well in his important book *Value of Values* (2007):

> When I look at any achievement I find that it is there because of certain opportunities I had, as well as because of my personal effort. I cannot claim to have created or commanded the opportunities; they were given to me. I happened to find myself in the right circumstances, and so I could grow and learn what I needed to learn. I met with the right person; happened to read the right book; I enjoyed the right company; someone came forward with the right guidance at the right time. There are so many factors behind an achievement. I cannot really say I created any of them. When I look at the facts, I must see that any achievement that I claim as mine is not due to my will or skill alone but it is due to certain things and opportunities that were provided to me.[12]

But there is a natural tendency to want to claim *personal* credit without acknowledging external factors. David Hawkins called this "the narcissistic core of the ego." It's just a natural part of who we are in an animalistic body. The ego distorts reality. When we examine accomplishments from a more holistic view of reality, pride starts to look silly. We're then left with *gratitude* as the only rational reaction to successes and a strong sense of obligation to use accomplishments to better the world in some way as a steward.

Let's look at examples from my life. Each of us could do this for our own unique circumstances, but I'll be the guinea pig here.

Question	Answer
In what area(s) have I succeeded?	I have experienced acedemic and professional successes (Princeton, Wall Street, Silicon Valley, and so on).
How would the narcissistic core of the ego interpret the successes?	I'm so smart. I consistently did better than so many people for years and succeeded at the highest level. I'm simply superior. I worked hard for these successes and deserve all the credit.
What is the reality of the situation?	Take it easy over there, ego. You've objectively had successes, but you were situated in very favorable conditions. You grew up in a family that gave you financial security and a top-notch education from a young age. Many of your relatives are naturally intelligent, so genetically you were working with a rigged deck. You were trained early on to have successful habits. Your high school gave you a better-than-normal chance of getting into Princeton. If you had done slightly worse in some of your high school classes, or had a few more bad days, maybe your grades wouldn't have been good enough to get into Princeton. The scales tipped in your favor at every turn. Once you got into Princeton, you were set up with better-than-normal odds of getting a Wall Street job, especially as a Division I tennis player. And from there you had access to stages you never would have had access to otherwise. From a higher-level perspective, if your parents hadn't met, you wouldn't have even been born. And if generations of their ancestors hadn't done everything exactly the way they did, your parents wouldn't have been born. And if the Big Bang hadn't caused every atom in the universe to interact in the exact way they did for the last ~14 billion years, none of your ancestors would have been born. Did you do any of that?

	From a spiritual perspective, maybe you chose your parents before birth and simply don't remember it. Therefore, you were set up with high odds of having these experiences by virtue of incarnating into your family. "Mark the Ego" didn't make that happen.
	Beyond that, who knows how our individual thoughts are generated? Think about every decision you ever made en route to your successes. How do you know how much of that comes from Mark the whirlpool versus the broader stream? Who is directing this ship?
	Yes, you worked extremely hard and have been laser-focused. But your individual efforts have played a tiny role in your accomplishments relative to the myriad factors involved practically, cosmically, and spiritually.
What would be the rational response to the success?	I've been fortunate to have experienced accomplishments and therefore feel gratitude for them. I can't take credit. I'm pleased with my efforts: I've given 100 percent effort with the deck I've been handed. And I feel a sense of responsibility to use my successes, skills, and resulting resources in whatever way best serves the whole because that's largely why I exist.

While conceitedness is clearly irrational, I'm not suggesting that we should downplay our abilities. Rather, I'm advocating for honesty. Sometimes we objectively perform well or have innate skills that are undeniably strong. Sometimes we are objectively "better" in some way. We shouldn't then employ a false humility and deny what we are. That would be out of alignment with truth and might cause us to miss opportunities. Nonconceit doesn't preclude self-honesty.

Furthermore, we could do the same exercise with our failures or "weaknesses." Our narcissistic ego tends to want to wallow in our failures and think, *Poor me, I'm so bad.* Upon close examination, we find that many factors beyond our immediate control led to the "failure." And therefore we are best served to approach it from a perspective of nonjudgmentalism. We treat the "negative" as an opportunity to have specific learning experiences that enable further Evolution of our consciousness. Failures and weaknesses aren't statements about our self-worth. "Regret" then becomes less rational, and we spend less time dwelling on the past.

In essence, all of this is advocating for brutally honest and objective assessments of everything.

Commitment

The One Mind model suggests that we are infinite, interconnected beings with amnesia. We are here, at least in part, to Evolve our consciousness. If we *fully* internalize these concepts, setting our life's compass to align with this direction becomes a top priority. All of our desires and struggles relate back to a desire for love, peace, and happiness—which leads us to the One Mind within us.

Why not go directly after it? David Hawkins said it well: **"Straight and narrow is the path. Waste no time."** The notion of *not wasting time* seems prescient given how many crises are emerging in the world today.

It took me a while to get to this level of commitment. I was only "sort of" on board in the early days. And then with further research and personal experiences, my conviction in this overall direction increased. Eventually one reaches a tipping point when one realizes that there's no such thing as being half-pregnant. For some people, it's a gradual progression. For other people who have a single, life-changing experience (for example, near-death experiences, meditation experiences, psychedelics, and so on), it's more of an overnight shift.

However, when we do arrive at this tipping point, commitment naturally enters the picture. Trying to uncover Truth and aligning with it becomes the only way to be in life. Adyashanti often calls this the "spiritual impulse," which isn't something the individual willingly generates. It comes from the infinite field of consciousness and pulls us. It pulls us to enact *its will* such that we are serving it.

A shift in priorities and lifestyles might result, but we're still human beings, living on Earth, with many responsibilities and relationships. I'm not suggesting that we drop all of those, live in a cave, and not care about daily living because it's too mundane. I'm instead advocating deep engagement in everyday activities, mundane or not. The activities themselves might not change, but they're performed from a new foundation. Then everything becomes full of meaning. We can see growth and learning in all aspects of life. In my own life, I feel much more engaged than I ever have before.

"Commitment" can take many forms. The yogic tradition lays out four general pathways that I think serve as helpful frameworks for a spiritual journey of any kind. Each of these pathways can have associated "spiritual practices" that can be performed daily, whether it's reading, meditating, chanting, doing yoga postures, or taking a certain approach to life. Basically anything you can think of can be put into one of these categories. What follows is my drastically simplified version of the four traditional pathways:

1. A focus on knowledge and wisdom *(Jnana Yoga)*
2. A focus on selfless service *(Karma Yoga)*
3. A focus on love for and devotion to the divine (One Mind) *(Bhakti Yoga)*
4. A focus on energetic aspects (for example, meditation, breathing exercises, physical yoga, qigong, and so on) *(Raja Yoga)*

A person typically starts with a focus on one of these areas, but eventually all four are naturally incorporated as one elevates one's consciousness and Evolves.

I'll relate my story as an example. In hindsight, I can see very clearly that I started on the *Jnana* (knowledge/wisdom) path—I was reading and listening to podcasts nonstop to learn. That path still continues, but through writing and other media, I feel that I'm doing a service to others *(Karma Yoga)*. And that desire to help others could be considered a form of love for the collective/One Mind *(Bhakti Yoga)*. Additionally, I've begun meditating much more and taking yoga classes (*Raja Yoga*). I'm constantly making sure my life is hitting all four areas.

This process can take time. I'm reminded of the Vedic analogy in which one dips a cloth into colored dye.[13] After the dye sits in the sun, the color fades but is still there somewhat. Then one dips the cloth back into the dye and lets it dry in the sun. The color fades again, but perhaps the color is slightly deeper because of the previous dip. And so on. Analogously, the cloth's coloring could be likened to the extent of our direct connection with the One Mind. With commitment to continued spiritual practice (in any of the four categories), we are regularly dipping our cloth back into the dye. Eventually, our normal state of being becomes increasingly tapped into the stream of consciousness.

One's "cloth dipping" method (that is, spiritual practice) likely varies by person. I've seen nothing to suggest that one path of spiritual growth is universally "superior." The intense desire to grow and Evolve consciousness is the key. It's all about where we set our compass.

Kabbalistic rabbi Aryeh Kaplan stated it well in his book *Jewish Meditation* (1985): "If you are willing to devote your life to continued growth, there are virtually no limits to the levels you can reach."[14]

Individuals who venture on this path may, over time, directly experience high states of consciousness, deep alignments with the One Mind. Sometimes these are called "enlightenment" or "awakening" experiences. If we're on the path, it's helpful to know what they are so we're well prepared. And furthermore, our understanding of these stages is an important step in the refinement of our compass. The next chapter dives into this topic.

CHAPTER 5

A PATH WELL TRAVELED

The Path

As I write this book, I'm still only a few years removed from my own initial spiritual awakening. Previously I would have regarded my current outlook as extremely esoteric. I think it's only considered to be that because our society is in a primitive state of development, very much out of alignment with reality. I think our future civilizations will look back at us in the way we look back on the time of cavemen/women. One day this stuff will become commonplace.

To my initial surprise, I've found that many people are—and have been—on this very path for thousands of years. You just have to know where to look. And with the advent of the internet and technology such as YouTube and podcasts, the information is readily available to basically anyone this very instant.

Along my journey, I've prioritized finding those who have already traveled the path. That way, I don't have to reinvent the wheel, can learn from other people's mistakes, and am able to avoid possible blind spots. The list of teachers mentioned in the previous chapter and Rick Archer's *Buddha at the Gas Pump* podcast have been go-to resources from afar (some of these individuals are no longer living). Additionally, I've worked with a number of people directly. It's helpful to be able to call on trusted advisers who provide valuable input and also reassure me that I'm not going crazy (or alternatively, to tell me that I've lost it—it happens!). I've built what I call an ever-growing and diverse "army of counselors."

Life can start to become quite mystical when one really sets out on this path. As I've begun to have personal mystical experiences on my journey, one of my advisers said to me, "We aren't in Kansas anymore, Mark, are we?"

I've seen it in my own life and in others': the more we open up to the One Mind view of reality, life changes around us in uncanny ways. As Rick Archer says, it's as if the universe perks up, proclaiming: "All right, guys, we've got a live one; let's give him or her some juice!" It definitely can feel that way.

At first this journey can seem daunting since it's a leap into the unknown. It's a new paradigm for living, which means leaving behind what is comfortable. We don't know where the journey will take us, and we simply surrender to the intelligence of the One Mind. As the thirteenth-century Sufi mystic Rumi said, "As you start to walk on the way, the way appears."

Awakening

In a recent conversation I had with a friend, he mentioned that he had gone on a hike with a prominent Silicon Valley tech executive. Both of them are on spiritual journeys, and the executive asked my friend, "So, are you awake?"

People toss around terms such as *awakening* and *enlightenment*, but often without proper context. As Archer often notes, these terms have a superlative, static connotation. I'd suggest that the terms can be more accurately thought of as a spectrum that never ends.

From my perspective, full enlightenment is complete union with the One Mind. We might call that *self-realization* (realizing our true nature, our True Self beyond our individuality). In fact, the term *yoga* explicitly means "union."

From the Absolute perspective, the One Mind is fully enlightened. And at some level, each of us *is* the One Mind. That's our highest identity. So that means we're all already enlightened.

Great! We can stop right there since there's nothing for us to attain. Right?

Not so fast.

At the Relative level, we are not the One Mind. We are the whirlpool. We are the individual sense organ of the infinite, one of the One Mind's vehicles of Relative experience. At that level, we are veiled from our true nature as the One Mind. And so you identify as a person as you read these words, not as an infinite field of consciousness that's beyond space and time.

So enlightenment at the Relative level is the extent to which we remove the veil. Let's return to our analogy where the One Mind can be likened to the sun. The clouds of our individual life (thoughts, conditioning, and so on) block the sun's rays from getting through. By removing the clouds, we allow in more rays of the One Mind so that we are more in union with it. The more rays we allow in, the more "enlightened" we are. At some point, we can theoretically approach 100 percent authenticity; we become pure consciousness—the One Mind—acting through this physical vehicle without obstruction. Then the vehicle can exercise its unique talents in full accordance with the One Mind and enact its intelligence.

This implies, as many spiritual teachers say, that enlightenment is not an attainment. Rather, it's a process of *subtraction*.

A further implication is that *each of us* is on the enlightenment spectrum, but simply at different stages. It doesn't matter if you're a murderer on death row or the Buddha, you have some degree of enlightenment. What varies is the number of "clouds."

"Awakenings" are noteworthy events that occur on this pathway to ultimate subtraction. They might be thought of as instances of rapid cloud removal. Some are incremental, whereas others are fast, intense, and earth-shattering. As Archer often says, "When the postman knows you're going to move, he tries to deliver all your mail."

When awakenings are significant, one's sense of identity can shift from being an ego/individual to feeling like pure consciousness (that is, the One Mind). But even after that initial realization, there are often many additional steps toward further awakening that occur. **It doesn't end with one flashy experience.** This is a common source of confusion.

Mechanisms of Awakening

A recurring theme in my research is that awakenings are often initiated by trauma or tragedy, often called a "dark night of the soul." As discussed in chapter 1, I had a dark night of the soul. Life felt meaningless and I was unhappy. But I had no idea it was a dark night of the soul at the time. I wasn't looking for spiritual awakening, but somehow it found me. I've learned that I'm not unique; this happens all the time. And as miserable and hopeless as I felt during that phase of my life, I had it way easier than some people.

It's almost as if when we're out of alignment with the One Mind (and out of alignment with why we incarnated in a body), the intelligent One Mind provides us with small nudges in our life. But if we don't listen and change our behavior, things get more drastic until we have no choice but to listen. We would be well served to listen to the cues early on to avoid undue suffering! With this notion in mind, we begin to pay close attention to what's out of alignment in our lives (what feels like "rat poison"). For this reason, Nisargadatta Maharaj said, "Life is the Supreme Guru." We learn from what life is presenting to us; hence, author Tony Robbins's often-quoted assertion: "Life happens *for* you, not *to* you."

Trauma isn't the only way to awaken. In her important book *When Spirit Leaps: Navigating the Process of Spiritual Awakening* (2018), transpersonal psychologist Dr. Bonnie Greenwell laid out a comprehensive list of "portals of awakening" she's seen in her career as an adviser to thousands of people going through awakenings. Examples include:[1]

- ○ Meditation

- ○ Movement (yoga, qigong, and so on)

- ○ Breathing

- ○ Energetic transmission from a guru or teacher

- ○ Body therapies (reiki, and so on)

- ○ Psychedelics

- ○ Shamanic initiations

- ○ Near-death experiences

- ○ Sexuality

Greenwell also cited cases of awakening that occur from what she called "spontaneous moments." Examples include dreaming, feeling strong love or devotion, reading a book about awakening, falling off a horse and hitting one's tailbone, sitting with a loved one who's dying, having food poisoning, walking home from school, giving birth, witnessing a beautiful sunset, going through surgeries, experiencing deep therapy sessions, surviving muggings, spending multiple hours with a biofeedback machine, visualization practices, and navigating the dying process.

The main point is: awakenings can be stimulated by many distinct events, and for each us, it's a different recipe. And who knows how karma plays into all of this? For example, if in a past life we removed many "clouds," do we then enter this life with a greater predisposition to have intense awakenings? Many believe this is the case—that we pick up where we left off in the previous life.

Even though predicting the timing, mechanism, and type of awakening experience might be impossible, we can perhaps improve our chances. "Spiritual practices"—like the ones Greenwell laid

out—might be such vehicles. They fall into the four yoga categories mentioned in the previous chapter under "commitment"
(briefly summarized as wisdom, selfless service, devotion, and
energy practices). As it's often stated on the *Buddha at the Gas
Pump* podcast, **"Awakening is an accident, but spiritual practice
can make you accident prone."**

Meditation

The most dramatically impactful practice I've personally experienced is the extended silent-meditation retreat. I've done one
six-night retreat and one five-night retreat so far. Those experiences seemed to have kick-started physical energy experiences
that I had never felt before and shifted reality in ways that are
difficult to describe. I have a new appreciation for meditation after
being resistant for years. And I've come to realize that doing *a
lot* of meditation is where the impact begins to be felt. Small bits
don't do that much (at least for me).

The popular notion that "meditation reduces anxiety"—while
true—is shortchanging something much more profound and
mystical. Meditation is an opportunity to "be still" (as Ramana
Maharshi phrased it), and not having any specific expectations
while aiming to connect with the One Mind within us. It's a
chance to remind ourselves who and what we are and sit in a quiet
state of surrender. And perhaps it even allows us to more easily
"download" some of the One Mind's intelligence. In Eastern traditions, some believe that meditation burns away karma. When
asked how this happens, Sri Sri Ravi Shankar responded, "It just
happens! It's natural. You keep anything on fire, it gets heated up,
and so meditation definitely erases the past impressions."[2]

When I meditate now, it's as if my battery is being charged with
energy—palpable energy that flows through the body. I also like
to keep in mind Rev. Bill McDonald's advice: If we add the feeling of love (giving and receiving) to any meditation practice, the
practice is automatically improved.

But everyone is different. I recommend trying out practices for
yourself to see what works best for you. I'm still exploring. I

hope this book has provided a starting point from which you can explore further.

A Precedent for Awakening

Once it hit me how important the topic of awakening is—and I realized I knew nothing about it—I recognized that there was only one thing left to do: study as many awakening cases as I could find. That's why I've been so drawn to the *Buddha at the Gas Pump* podcast. It's an opportunity to hear how hundreds of ordinary people have gone through this life-changing process. The implication of the show's title is that you could be at the gas station, and unbeknownst to you, the person pumping gas next to you is at the enlightenment level of the Buddha.

As podcast host Rick Archer says, there's an "epidemic" of spiritual awakening occurring now. Part of the reason he started the show in the first place was to demonstrate that this underground movement is truly happening; it's not just a few one-off experiences that we can dismiss as anomalies or people going crazy. You don't hear about it on the news, but there's an upwelling of such cases occurring all over the world.

However, there are a number of famous cases throughout history, both in the East and the West. One noteworthy compilation is Dr. Craig Pearson's book *The Supreme Awakening: Experiences of Enlightenment Throughout Time—and How You Can Cultivate Them* (2013). He presents accounts of historical figures who have reached various stages of awakening. Included in the book with each example are relevant quotations suggestive of a certain "level" of awakening.

Examples in the book include Plato; the Buddha; Jesus Christ; Marcus Aurelius; St. Augustine; Rumi; Meister Eckhart; St. Teresa of Avila; St. John of the Cross; Shankara; Ralph Waldo Emerson; Alfred, Lord Tennyson; Henry David Thoreau; Walt Whitman; Emily Dickinson; Helen Keller; Howard Thurman; Anwar el Sadat; Plotinus; Rabbi Abraham Isaac Kook; Black Elk (Hehaka Sapa); Dante Alighieri; Leo Tolstoy; William Wordsworth; Jane Goodall; Laozi; and many others.

Typical Awakening Experiences

So, what exactly are awakening experiences like? We touched on this in chapter 2 when discussing common themes in spiritually transformative experiences. There's a universal quality in which people—from all over the world in different time periods—have and still are tapping into the same reality—the One Mind.

Bonnie Greenwell summarized the typical characteristics: "Awareness and consciousness are suddenly clear and expansive, undisturbed, and undivided by thought. The experience may be accompanied by great insight, ecstatic bliss, or a mystical infusion of light, love, and vision. **Awakening can strike us like a bolt of energy or it may gently unravel through years of seeking truth**"[3] [emphasis added].

She offered an example of a librarian's awakening while meditating at a retreat:

> There was a sudden seeing of complete, white, brilliant stillness everywhere. There was no "me" thinking—there was nothing but white illumination and it was complete love. It was spacious, unbounded, undifferentiated, endless, and pure love. I became aware that everything is of love and there is nothing other than that. Also, within this brilliance arose little blobs of worry, angst, and thoughts. I immediately knew that every thought, worry, angst was only that—and nothing else. They just appeared and fell away, and were without substance in this complete and pure brilliance. There was only joy, awe, and amazement.[4]

Ego Death and Awakening

Awakening experiences can include a sense that the individual self is dying. The identification with the whirlpool goes away, and one merges with the stream, which can feel terrifying. It's a death—the death of a false identify we've all been clinging to. There are many, many examples of this, but I'll mention a few cases here.

Spiritual teacher Adyashanti's first awakening experience has stuck with me since I first heard about it. He was an avid meditator in the Zen Buddhist tradition and would sometimes wake up at 4 A.M. and meditate until midnight. He so badly wanted to have an awakening breakthrough. One day, at age twenty-five, he sat down to meditate and was particularly determined to achieve an elevated state. But not long into his meditation, he became defeated (intellectually, physically, emotionally, and energetically). He thought he couldn't do it and that the awakening experience would never happen. But as soon as he had this sense of being defeated, he had what he calls "an immense opening." He started breathing heavily, and his heart was beating rapidly. As he stated in an interview:[5]

> It felt like my heart was going to explode.... I just had a thought: *This is gonna kill me.* And the next thought I had when I knew it was gonna kill me was: *If that's what it takes, let's get it over with.* And it wasn't like a courageous thought, it wasn't a masculine, fearless *Okay I'll do it;* it was just a very simple, *Okay I'll die. I'll die right here, right now. I'll let my heart explode. I won't move off this cushion. I won't go anywhere.* And I really thought I was gonna die. 100%. As soon as I thought, *Okay I'll die now* all the sudden I was instantly in a different... dimension. I wasn't aware of my body anymore... and I wasn't even there anymore. It was just this infinite open black space.... And after a while, what I could feel is like there were these insights coming, almost being downloaded through the top of my head into my body, like a computer program was being downloaded.... They were coming like 100 a second.... If you can imagine "ah-ha's" going off like a popcorn popper.... And then... slowly... I became more aware of my body. The energy had settled... and after a while I was just sitting there, totally normal.... No high, no low.... I think, *Well, there's nothing left to do. I guess I'll get up.*

He then bowed at his Buddha statue and started laughing hysterically. He looked at the Buddha and said, "I've been chasing you for years!" He commented, "And I could see that which I had been chasing was this that was chasing it.... **I knew that what I had been chasing was what I am.** And that's where the laughter came from" [emphasis added].

This statement was profound. His initial "seeking" of enlightenment felt like searching for something outside of himself. But when he had the awakening, he felt viscerally that he *was* the One Mind. He realized that he'd been chasing himself. It was so obvious and yet so elusive. This is the key revelation often reported—people get an experiential understanding of this concept rather than an intellectual one.

Previously I had read about many of these experiences and couldn't quite grasp what people were saying. It wasn't until going through something similar myself that I at least partially "got it." Twice during recent meditations I was overtaken by an overwhelmingly blissful energy that is difficult to describe. Both times the energy arose while I was meditating with the specific intention of surrendering. I had the feeling, without saying it: *There's really nothing that I need or want from the physical world anymore. I can just be a vessel and serve.* The energy entered without my having any control over it. It just started flooding my body. It had an otherworldly, spiritual quality to it that I had never before encountered. And the pleasure was far beyond anything I had felt in any aspect of physical life. But as the energy overtook my body, my heart felt like it was beating out of my chest. My body completely panicked, but my mind was trying to surrender. There was a silent but clear feeling that I was going to disappear and no longer exist. At that point, my body entered a state of terror and physically shut off the energy. The energy subsided but sometimes comes back in smaller doses during meditations.

I mention my story here not to brag about it, but rather to provide a sense of why this topic is of such high interest to me. When this stuff happens to you, it's all of a sudden not theoretical! And on this path, experiences like this are common.

Another awakening experience I've enjoyed studying is that of David Hawkins. He experienced fear, but like Adyashanti (and unlike me, so far), he went past it. He described the occurrence:

> The last confrontation arises unannounced, and therefore, it is never too early to prepare. It can happen to a seeming novice, a "spiritual idiot," or even an atheist at the point of death.... The "final moment" opens up in a split instant as an overwhelming illumination, realization, and presentation. The last step can be the consequence of the elimination of all that previously stood in the way by virtue of diligent spiritual practice. There are often preliminary warning flashes of advanced insight... sudden unbidden moments of absolute stillness and peace in which time stops.... The underpinnings of the ego are its illusion that it is a separate self.... When these structures are transcended, the ego brings up its last reserves. These consist of the threat of death.... When this arises it becomes rapidly clear that one is now forced to make a decision and choose.

The "choice" here is whether one is willing to "die" as an individual self. That's the choice Adyashanti made. Hawkins continued:

> By invitation and surrender, death becomes an experiential reality. It can be fearful and intimidating for a brief moment. It is not like the physical deaths that occurred in previous incarnations when one left the body with great relief. This is actually the first and also the last time that real death can be experienced.... With the courage of conviction... one surrenders to the plunge. For a few moments, the last great fear erupts and one experiences what it really means to die completely as the great door swings open to the Splendor, beyond all comprehension. The Presence reveals that the Infinite Splendor is actually one's own Self. Innate is the knowingness that one's reality is beyond all lifetimes, beyond all universes, total and complete. One knows the Allness because one is the Totality. There is nothing left to know nor anyone

to know about it. The Presence obliterates all but Reality. One is "home at last."[6]

I am emphasizing these stories because it's helpful to know that this is objectively what happens. However, my concern is that readers will latch on to the extraordinary nature of these temporary experiences. The real test in life isn't whether we can have an experience and then tell people how enlightened we are. That's "spiritual ego," which we'll discuss again soon. Rather, the test is whether we can *embody and stabilize* the awakened state in everyday living to improve our own lives and those of others.

When we fully embody this state, we live each day in full alignment with the One Mind. Some call this stabilization state "liberation," a concept often referenced in Eastern religions (*moksha* in Hinduism or *nirvana* in Buddhism). As Bonnie Greenwell put it, "We become liberated from our old identifications, compulsions, demands, and suffering."[7]

That said, part of the embodiment process is maintaining *balance* in our lives. Rick Archer gives important advice:

> It can be as challenging to integrate spiritual experience as it is to achieve it. If you have an intense spiritual practice, and/or are having powerful inner experiences, unless you are engaged in dynamic, challenging outer activity, you can get so absorbed in your subjective world that your engagement in the "ordinary" world may appear bizarre to its inhabitants. Spiritual seekers often want to evolve as quickly as possible. It may be tempting to join an ashram, meditate umpteen hours a day, go on a long fast, become celibate etc., and these things may have their time and place, but if you take them to extremes and don't maintain balance and integrate your experiences, you can end up in a spiritual and psychological cul-de-sac.[8]

Kundalini Energy

The energy referenced in the previous section has a name. It's often called "kundalini energy," which is stored at the base of the spine. Bonnie Greenwell commented: "Skeptics consider it imaginary. Few who have not experienced it believe in its existence. All who have experienced it know it as a mystery and a profound life-altering experience."[9] I can attest that the energy is simply unmistakable. But I can understand how someone might be skeptical in the absence of having had a direct experience.[10]

The science of how kundalini activates and flows in the body during awakenings is fascinating. Dr. Joan Harrigan's lengthy books *Kundalini Vidya: The Science of Spiritual Transformation* (2002) and *Stories of Spiritual Transformation: The Fulfillment of Kundalini Process—Modern Seekers, Ancient Teachings* (2017) offer in-depth explanations as to how this works and what the results have been during modern spiritual awakenings. It's been mind-blowing to learn that these complex systems have been understood by spiritual traditions for thousands of years, including how the energy flows and can become blocked and then corrected. I would imagine that at some point, when these awakening experiences become more mainstream, our modern science and medicine will catch up.

The kundalini process induces similar symptoms across the board. It's helpful to know about this before the energy begins to move, so you're not overly alarmed if it happens suddenly. And if you're having it and don't know what it is, maybe the following explanation will help you contextualize your own experiences.

Most Western doctors don't know what to do with this stuff. Fortunately, transpersonal experts exist who can guide individuals through this positive and life-changing process, such as Bonnie Greenwell, Joan Harrigan, and Dorothy Walters (and discussed very often in *Buddha at the Gas Pump* episodes). In Greenwell's book *The Kundalini Guide* (2014), she reviewed seven typical categories of symptoms she saw in clients having kundalini awakenings, summarized here:[11]

1. *Pranic activity or kriyas:* "Kriyas are involuntary body movements, shaking, vibrations, jerking, and the sensation of electricity, tingling, or rushes of energy flooding the body.... Sometimes people wake up in the middle of the night and their body spontaneously moves into a yoga posture, or performs a gesture with their hands that is used in yoga to bring peace or focus." Kriyas can be gentle and blissful but sometimes can feel uncomfortable. They can also occur as more general releases outside of a kundalini energy flow process.

2. *Other energy phenomena:* The mind can produce images of other worlds or images of a life that is not their current one: people might also hear sounds of chanting, Sanskrit words and tones, bees buzzing, and kettle drums beating. One of Greenwell's clients was hearing Sanskrit, not even knowing what language it was until a friend invited her to an Indian event where she recognized the songs. Some people feel a need to purify their home, engage in a ritual, or go on a pilgrimage.

3. *Physiological problems:* Kundalini can cause illnesses, and it can bring up something hiding beneath the surface. People can become overstimulated, which then leads to exhaustion. Insomnia is commonly reported. Eating patterns can change such that people lose interest in certain foods or drugs and alcohol. Sexual patterns can change as well.

4. *Psychological and emotional upheavals:* Unresolved emotional issues can come up along with mood swings, anxiety, and other issues. While unpleasant, one should keep in mind that this is all part of a clearing process.

5. *Extrasensory experiences:* Acute vision and hearing is reported, such as seeing ants crawling up a tree or hearing what's happening in another room. More commonly, people report extraordinary visual images (lights and symbols). Sometimes people report a radical change in their environment, such as being in another historical time. There are also occasional reports of hearing a voice

or smelling incense.

6. *Parapsychological experiences:* Psychic capacities are activated, such as remote viewing (seeing at a distance with the mind), precognition (knowing the future before it happens), and psychokinesis (impacting matter with the mind alone). Unusual synchronicities are reported, as well as an awareness of auras and electrical sensitivity. (Electrical sensitivity is also commonly reported in people after having near-death experiences. Often their watches and appliances break, and even street lights flash strangely when they walk by.)

7. *Samadhi or Satori experiences:* These are mystical states of unity, peace, light, and energy: "a glimpse of the truth." Hawkins's description is particularly pointed: "It was necessary to stop the habitual practice of meditating for an hour in the morning and then again before dinner because it would intensify the bliss to such an extent that it was not possible to function."[12]

Before we move on, I feel obligated to offer a warning. Some practices, such as targeted breathing methods, can artificially and prematurely activate kundalini in the body. I've been told by experts repeatedly, "Be careful!" If the body isn't ready, the energy can be physically dangerous. If you're working with a teacher on breathing methods—or anything related to kundalini activation—be sure to ask about the risks.

CHAPTER 6

CAUTIONS

It's Not All Roses

I've alluded to some of the challenging aspects of the spiritual awakening process. While it can ultimately lead to incredible experiences and an overall improved life with a great sense of meaning, the path to get there can be psychologically bumpy.

I've mentioned "dark nights of the soul" and psychological upheaval that can occur during the kundalini process. Even Mother Teresa, who was an undeniably elevated being, experienced periods of great difficulty. Her several decades of "darkness" included a "terrible sense of loss" and a feeling that "God does not want me.... The torture and pain I can't explain."[1] Additionally, Sufi master Irina Tweedie stated, "The ego does not die with laughter and caresses; it must be chased with sorrow and drowned with tears."[2] One of the primary challenges is around the "stripping... of habitual affections and attachments to the old self," as stated by author Mirabai Starr. Dr. Mariana Caplan, a psychotherapist, summarized it well: "Some who aspire to great spiritual heights will need to endure a profound degree of suffering."[3]

The challenges don't necessarily end there. I've been stunned to learn how many pitfalls exist on the spiritual path. We will be exploring these areas so that they can be adequately incorporated into a comprehensive life compass.

Along these lines, Caplan's seminal book, *Eyes Wide Open: Cultivating Discernment on the Spiritual Path* (2009), could be considered required reading for anyone endeavoring to focus on spiritual Evolution. The sooner one becomes familiar with the principles laid out in that book, the more one can avoid unnecessary distractions and suffering. In the next few sections, I will summarize many themes she and other spiritual counselors have often explored.

Premature Immaculation

Flashy awakening experiences are dramatic eye-openers indicating that we're heading in the right direction. But we have to be careful not to get ahead of ourselves and automatically think we're so special just because we've had an amazing experience (and that goes for me too!). Flashy experiences don't suddenly make you the next great saint overnight. This is apparently a common problem and a theme that emerges in some of the *Buddha at the Gas Pump* episodes. Rick Archer has often warned against this "premature immaculation."

Mariana Caplan similarly cautioned that powerful awakening experiences can be just the start of someone's journey rather than the end. If the One Mind is constantly Evolving, and each of us is part of the One Mind, is it even possible to say that *anyone* is "done"? As Caplan said, "Rather than holding on to an imagined concept of enlightenment as a product that can be attained, it may be more helpful to consider spiritual awakening as an endless process of progressively deeper levels of integration."[4]

So if you start having amazing experiences of unity and bliss, enjoy them, process them, and be thankful that you're on a path forward. Just be mindful not to make faulty assumptions about your development. And furthermore, beware of idolizing others (particularly teachers and gurus) who have had those experiences (more on that coming up).

Spiritual Ego and Spiritual Materialism

Flashy awakening experiences can create an ego that is actually counter to the intent of the spiritual endeavor in the first place. Judging our own experiences against those of others can be counterproductive and distracting. Each of us is wired differently, and awakening manifests differently for different people. And we don't know how karma and other factors play into it. Developing what Mariana Caplan called "spiritual superiority" is dangerous.[5] But, like all emotions, if we start feeling emotions of superiority, we can simply allow them to come in them without resistance. We let the emotions run their course, while not acting on them inappropriately. Over time, the unwanted emotions should run out.

Caplan stated it well: "We all resist seeing the ways in which we deceive ourselves on the spiritual path. It is an embarrassment to the ego."[6] It can be helpful to have third-party advisers who can give you fresh opinions and let you know if ego is creeping in. Even spiritual teachers whom we admire could probably benefit from this. As mentioned on the *Buddha at the Gas Pump* podcast, Miranda Macpherson (a spiritual teacher herself) goes on retreats with multiple spiritual teachers every year. That sanity check helps us avoid ego traps and actually assists with Evolution.

Similarly, the goal of the spiritual process isn't simply to accumulate flashy experiences (sometimes called "spiritual materialism"). If we overemphasize the experiences, we can get caught in an ego trap of "*me* and *my* experiences" rather than consider how we can contribute to the world. Ultimately, we aspire to be the best people we can be, and awakenings are often just part of the development process. I've stated it before but will mention it again: if we can't hold the energy of those experiences in our daily living, then what good are they?

Lines of Development and Spiritual Bypass

Author Ken Wilber's work on "lines of development" is critical. The basic idea is that people develop in distinct ways across different aspects of their being. For example, someone may be incredibly awakened, psychic, and capable of perceiving interdimensional realities, but at the same time have serious interpersonal problems. Wilber's research suggests that lines of development are largely independent, which leads to mismatches. **So it's possible for someone to be seemingly "awake," but as a teacher acts like an "egocentric fascist"** (as Wilber put it).[7]

He summarizes by describing three primary ways in which we develop:

1. Waking up (that is, awakening)
2. Cleaning up
3. Growing up

Wilber contends that many spiritual traditions adequately focus on the waking-up process. However, those traditions are ancient and lack integration with modern psychotherapy. For instance, the notion that we have suppressed emotions is a relatively new development in psychology and isn't part of most spiritual awakening teachings. Wilber sees an exciting opportunity because, for the first time in history, we can bring together waking up, cleaning up, and growing up into one package. **Therefore, if we're on the path toward Evolving ourselves, it's useful to engage in a combination of spiritual practices *and* some form of traditional psychotherapy.**

Ignoring "shadows" we carry is a form of *spiritual bypass* that can be damaging to our process. It's best to address those early so they don't bubble up unexpectedly and stunt our progress. But we don't want to become overly fixated on "not bypassing" and shut ourselves off from having spiritual experiences. We just need to give ourselves room to work through our deep-seated issues. We're all human, so we all have them!

Wilber offered an important warning in an episode of *Buddha at the Gas Pump*:

> If you're going to join a [spiritual] path and it doesn't include something on all three of those [that is, waking

up, cleaning up, and growing up], you're gonna get
fucked. You are gonna end up in a bad, bad way because
one of these is gonna go sour. Because you're completely
unaware of it and it's gonna be operating in the nasty
way that it does, and at some point it's gonna bite you
and you're gonna be extremely unhappy.... [It's] incredibly common: you have people that are fairly... spiritually
evolved, and then they get into financial, sexual, power
screw-ups because they're at a fairly low level of growing up.... In spiritual practice we train only for how to
transcend the individual self. We don't train for how do
you actually help grow that individual self, even though
you're gonna transcend it and... realize the reality of
ultimate oneness. You still have to express that oneness
through whatever self you have. And if your self is at a
low egocentric stage of development, and if you have
a massive shadow element and you try to express an
otherwise authentic enlightenment, it's going to be a
disaster.[8]

Our "shadow" is the part of us that still needs healing and is unresolved. It's the part of us we know we could improve. Unresolved
trauma has to be dealt with. It could be from this life, but it could
also come from ancestors. For instance, the study of epigenetics
suggests that *trauma might be passed down through genes.*[9] Our
trauma might not even belong to "us"! Furthermore, who knows
what past-life traumas we carry?

So handling our shadow is simply part of the Evolutionary process. We clear out the gunk so that we can be pure vessels. Spiritual
practices can help the process, but we should be aware that they
might unearth shadow elements that then need to be processed.
As Mariana Caplan advised: "Immersing ourselves in the puri-
fying power of spiritual practice often surfaces our wounds more
readily, and with greater intensity, than maintaining a life filled
with psychological buffers and staying within the confines of conventional paradigms of mainstream culture."[10]

Teachers

Ken Wilber's lines of development aren't just critical for our own growth, but they're essential when evaluating potential teachers, gurus, or coaches in the awakening process. We should be mindful of whether the person is at a high level of development *in all three areas*, or if, on the other hand, there is a dangerous imbalance.

The value of teachers is that they're a bit farther along in the journey and can guide and provide feedback. Sometimes they recommend specific practices that will help advance our process. We just have to be *really careful* whom we engage.

Caplan offered great advice on this topic: "It is important to acknowledge the pervasiveness of scandalous behavior on the spiritual path. While life-threatening instances of abuse do occur, they are the exception. Most of the infringements are milder— and therefore less visible—including psychological, financial, and sexual coercion. Even more common is the phenomenon of spiritual mediocrity, which occurs when unprepared individuals prematurely place themselves, or are placed in, positions of power they are not equipped to handle."[11]

She also noted, "Celibate yogis, swamis, rinpoches, and tulkus have come from the East to teach and quickly found themselves sleeping with attractive young students while professing celibacy."[12]

Many teachers have succumbed to the allure of power, money, and sex. The more I learn, the more shocked I am to hear about scandals in this arena because they seem so contrary to the spiritual intent. A popular example was that of spiritual teacher Osho, chronicled in the Netflix documentary *Wild Wild Country* (2018). In the 1980s, Osho transformed a 65,000-acre Oregon ranch into a city, including an airport. The commune fell apart after countless examples of criminal activity.

Just because a teacher is awakened in some way doesn't mean that he or she should be allowed to get away with otherwise unsavory behavior. There can be a tendency to sweep negative traits under the rug and give teachers the benefit of the doubt. That's how cults can and do form. It's dangerous to give people a "pass" for unacceptable behavior just because they're "spiritual."

The Association for Spiritual Integrity (ASI) is a charity founded in 2018 by Craig Holliday, Jac O'Keeffe, and Rick Archer that aims to prevent scandals and otherwise detrimental behavior in both teachers and students. The student is sometimes at fault just as much as the teacher. ASI published a Code of Ethics and Good Practice for Spiritual Leaders, Teachers, and Guides, as well as Student Guidelines. Anyone who is currently in, or aspires to be, in a student/teacher relationship would be well served to read these documents at https://www.spiritual-integrity.org/. Along these lines, I also recommend Mariana Caplan's book *Halfway Up the Mountain: The Error of Premature Enlightenment Claims* (1999).

Supernatural Powers

As mentioned earlier, a typical symptom of the kundalini process is enhanced psychic abilities. In the yogic tradition, they're called *siddhis*, but they're also mentioned by different names in the Catholic, Islamic, Jewish, and Tibetan Buddhist traditions. There have been extraordinary cases reported throughout history including levitation, appearing in two places at the same time (bilocation), instances when the body does not decay as normal after death (incorruption), making material objects appear out of nowhere (physical mediumship), and other seemingly magical concepts.[13]

As Paramahansa Yogananda's teacher Swami Sri Yukteswar put it: "There are subtler laws [than natural laws] ruling the realms of consciousness which can be known only through the inner science of yoga. The hidden spiritual planes also have their natural and lawful principles of operation. It is not the physical scientist but the fully self-realized master who comprehends the true nature of matter."[14]

Each case needs to be examined on its own merits, but in some instances there is apparently strongly corroborated evidence (for instance, see Dr. Dean Radin's book *Supernormal: Science, Yoga, and the Evidence for Extraordinary Psychic Abilities* [2013]).

I don't mention this topic so that we can debate the reality of crazy-sounding phenomena. Rather, it's a warning that on this path we need to be careful of being seduced by glamorous phenomena. We shouldn't automatically worship someone (or ourselves) just because seeming powers arise. Perhaps they can be helpful, but they can also be potential distractions from our Evolutionary path if they are misused or lead to narcissistic egocentrism. As David Hawkins often said, glamorous phenomena appeal to our "inner child." Just because someone can do miraculous things doesn't necessarily mean he or she has "cleaned up" or "grown up." We must remain discerning and remember what our overall compass is at all times.

Channeling

Along these lines, discernment is critical when encountering the phenomenon of "channeling." In my opinion, channeling is likely a real phenomenon, but it's important to beware of possible manipulation or other ill-intentioned traps.

Channeling is a form of trance mediumship in which the medium ("channeler") becomes a vehicle for a nonphysical being who literally speaks through the vocal chords of the medium. The "being" is typically in another dimension and isn't just someone's deceased relative. There can be all sorts of allegedly "advanced" beings that we can't see with our eyes. In striking cases, the channeler speaks in a completely different manner than normal. Tone, facial expressions, speaking cadence, and general presence can completely change. It's sort of like "being possessed," except the channeler is typically willing to allow in the being's consciousness.

Of course our knee-jerk reaction to this might be: "They're good actors," or "They have some kind of mental disorder." This appears not to be the case, however. Helané Wahbeh, director of research at the Institute of Noetic Sciences, studies channelers scientifically. She hooks them up to EEG and EKG machines, and the evidence suggests that they aren't faking it. Their symptoms are not in alignment with known pathological disorders, and the individuals tend to be high functioning and hold normal jobs.

Additionally, Wahbeh has found that when channelers are channeling, they exert some kind of physical energy that has a tangible effect on the surrounding environment. She tested this by placing a random-number generator in the corner of a room and measuring the flow of "1s" and "0s" when channeling was occurring versus when it wasn't (controlling for other variables). When no one was channeling, the machines did their job of spitting out 1s and 0s in a random fashion. As expected, there were roughly 50 percent 1s and 50 percent 0s. However, when channeling was occurring in the room, the machines behaved statistically nonrandomly. (For more, see *A mixed methods phenomenological and exploratory study of channeling* [2018] by Wahbeh et al.). As Wahbeh told me when I interviewed her, "What's so remarkable about this is it's an objective measure that's showing a shift in the environment while we had many, many aspects of the situation controlled."[15]

All of this suggests that channelers are doing something real. But I will admit that it's hard to believe it unless you've seen it yourself. I can say from personal experience with a number of channelers that something is definitely going on.

We have to wonder if some of the great historical religious figures and "prophets" were channeling, and others simply weren't understanding what was happening. More recently, many well-known channelers have emerged, such as Edgar Cayce, Jane Roberts (channeling a being identifying as "Seth"), J. Z. Knight (channeling a being identifying as "Ramtha"), Abraham-Hicks, Paul Selig, and many others. I recommend Kevin Moore's show *They Call Us Channelers* (available on YouTube) if you'd like to learn more and watch channelers in action.

The danger here is that it's easy to become "Wow"-ed by channelers. After getting over the initial shock, there's a tendency to almost worship the intelligent being speaking through the channeler. We could easily think, *I should listen to whatever this being is saying!* And sometimes, the beings are in fact advanced and helpful. Many of them talk about the Evolution of consciousness on the planet and offer useful advice. However, other times they're egocentric, manipulative, and even dangerous. In the same way we have "good" and "evil" people in the physical world, apparently it's the same in the nonphysical world. There's a spectrum.

In her episode on *Buddha at the Gas Pump*, channeler Gina Lake spoke about how she was manipulated by beings she was channeling. She thought they were benevolent, so when they told her she should quit her job, she listened. The decision turned out to be ill advised. The beings began laughing and told her she should kill herself. Lake commented, "I wasn't grounded enough to reach higher level beings all the time, and so I started getting... tricked.... At the time I couldn't discriminate between the higher level beings I was talking to sometimes and... lower... beings that I was talking to, so it was very confusing.... It was very devastating to realize that I had been... deceived."[16]

If we decide to explore channelers, we need to ask ourselves why we're doing it in the first place. A high-level channeler who appears to have little ego investment might help us learn something for our Evolutionary process. But if our interest comes from the curious "inner child" who is lured by glamour, that's when we're in danger of being distracted or even manipulated.

The Seduction of Power

David Hawkins told a story that I think is emblematic of temptations we all face in life, but perhaps even more so on the spiritual Evolutionary path. As he was elevating to higher and higher states, and continuing to surrender, he was faced with a proposition (from an apparently nonphysical source). The thought came to him: *Now that you realize you're beyond all karma, all power belongs to you. Own it. Claim it.* He reasoned that this didn't make sense, since he *was* everything (as the One Mind). So why would he have needed power over others? He therefore rejected the temptation and claimed that it unlocked an important step in his awakening process.

Similarly, the story of the Buddha included his being tempted by the nonphysical being "Mara" (a "demon"). And the story of Jesus Christ includes his being tempted by the "Devil" following his forty-day fast in the desert. Even Patanjali's *Yoga Sutras* warn against temptations from celestial beings and emphasize the importance of ignoring them.[17]

Whether or not one believes these specific accounts is almost irrelevant. What's most important is the theme, which I think *is* relevant. While on the Evolutionary path, temptations will inevitably arise. We should be mindful of the allure of power in advance so that it doesn't catch us when we least expect it.

Mistakes on the Evolutionary path are inevitable, but the consequences can be severe. Hawkins often talked about "responsibility versus culpability." If we accidentally hit someone while driving our car, we're responsible but not culpable. But if we possess spiritual knowledge and act poorly, we are both responsible *and* culpable. Regarding cases like these, I often hear the hypothesis that the consequences of our mistakes have more severe implications for our karma (high culpability).

For example, psychologist and Sufi sheikh Robert Frager said that in Sufism, a teacher's mistake to a student can carry double the karmic consequences of a normal mistake. And if the student goes on to teach that mistake to another student, the original teacher's karmic consequences are quadrupled, and so on.[18] The theme here is the old adage: "With great power comes great responsibility." We have to be even more mindful than usual of how we act and what information we transmit. This attitude further encourages humility.

Psychedelics

Psychedelics were mentioned in the previous chapter as a portal to spiritual awakening. Given how much attention they seem to be getting everywhere, they deserve discussion. In addition to stimulating awakenings, they can help us quickly "clean up" by unearthing emotional baggage and shadow elements.

A number of promising results have been reported. For example, in a Johns Hopkins study using psilocybin (the active psychedelic in "magic mushrooms"), six months after being treated with the substance, 80 percent of the study's terminally ill cancer patients reported experiencing less anxiety and depression about the prospect of dying. Two-thirds of them described their experience on the substance as one of their top-five most meaningful life experiences.[19] Johns Hopkins now has a center devoted to psychedelics research.

Additionally, MDMA (sometimes called "ecstasy") is at the phase 3 clinical trial stage for treating post-traumatic stress disorder after many favorable results. Others are using the Peruvian plant mixture "ayahuasca" (containing the psychedelic DMT) to clear trauma. And ketamine is now being allowed for depression treatment.

My opinion is that the world is in the midst of a mega-mental health crisis. I see that as ultimately stemming from our vast misunderstandings about life and reality. Conventional psychiatric drugs aren't solving the root issues, whereas in some cases psychedelics unquestionably work at a deeper level—especially when the "trip" is integrated with an experienced therapist. There seem to be benefits, no doubt.

But at the same time, I do hear horror stories about people who either have traumatic experiences or don't come out of their altered states. So in addition to being illegal in most countries, there are real health risks.

From the standpoint of spiritual growth, psychedelics can also exacerbate problems experienced during the kundalini process. Bonnie Greenwell counseled: "I advise anyone having difficulties with kundalini energy or invasive images and thoughts to avoid all intoxicants and mind-altering substances. Whenever overactivated, we need to direct attention to simply grounding ourselves by being in nature, eating healthy and having quiet time. A sign of a stable spiritual awakening is the ability to live in the moment with what *is* rather than grasping for the next big experience"[20] [emphasis in original].

There's another consideration that I don't hear discussed much: the consciousness of the plant (for plant-based psychedelics). For example, users of ayahuasca often talk about "Mother Ayahuasca" as if it's a distinct personality and consciousness that becomes activated in the user. The plant consciousness seems to have its own agenda when it's consumed. For some people, the benefits are strikingly obvious.

But I think we should be mindful of introducing "plant consciousness" into our system without truly knowing or understanding the plant's motives. It's reasonable to guess that the plant would have beneficent intentions, but we simply don't know all the facts. Does the plant consciousness linger and drive people's behavior after they take the psychedelic? If so, in what direction is the behavior driven? Does the motive of the plant consciousness differ by plant species? Does it differ within plant species?

By now you're probably wondering: Can plants even have consciousness? Let me clarify here that I'm not talking about the panpsychist view that plants *have* consciousness. Rather, I do think it's possible that the universal consciousness—the One Mind—could theoretically express itself through the vehicle of a plant. Remember: if consciousness doesn't come from the brain, then a brain isn't needed for something to have a conscious experience.

And, in fact, there's preliminary evidence beyond psychedelic accounts that suggests that consciousness *does* work through plants. The evidence came from Cleve Backster beginning in 1966. Backster was one of the world's best lie detectors, having done work for the US government and the CIA during World War II. While watering his plants during a coffee break one day, he wondered what would happen if he hooked up a plant to a lie detector (known as a polygraph). He tried to think of ways to make the polygraph spike by "threatening" the plant. He recounted what happened while standing fifteen feet away from the plant:

> My thought and intent was: "I'm going to burn that leaf!" The very moment the imagery of burning the leaf entered my mind, the polygraph recording pen moved rapidly to the top of the chart! No words were spoken, no touching the plant, no lighting of matches, just my

clear intention to burn the leaf. The plant recording showed dramatic excitation. To me this was a powerful, high quality observation.... I then thought, "Gee, it's as though this plant read my mind!" I left the room and went to my secretary's desk to get some matches.... When I returned the plant was still showing highly visible reactions.... After returning the matches to my secretary's desk, the tracing returned to the calmness displayed prior to the original decision to burn the electroded leaf.[21]

Backster devoted the rest of his life to studying this phenomenon and got many positive results. He even found that plants were attuned to human emotions (and that human cells were attuned to the host's emotions even when the cells were not in the human's body!). His book *Primary Perception: Biocommunications with Plants, Living Foods, and Human Cells* (2003) documents his years of experimentation.

Much more research is needed before we can reach definitive conclusions. But the implications for plant-based psychedelics—and of course for our thinking about the environment generally—are significant.

Overall, psychedelics feel a bit like the "Wild West" currently. Indigenous cultures have been using them for a long time, so we could benefit from learning more about their findings. Hopefully, continued research will be permitted and funded so that we can better understand the benefits and risks, and how we can safely use (or should avoid) psychedelics during a spiritual journey.

CHAPTER 7

RIGHT-SIDE-UP LIVING

A Compass Aligned with Reality

We have established a view of reality that turns the totally unproven physicalist worldview on its head. Consciousness doesn't come from the brain, as physicalists wrongly assume through a massive leap of faith; rather, consciousness is the basis of reality, and the physical universe emerges within it. Instead of viewing ourselves as temporary biological machines in a meaningless universe, we're something much more, and life is much more purposeful. Systems theorist and philosopher Dr. Ervin Laszlo summed it up well: "We are an infinite consciousness associated with a finite body."[1] Our society is overlooking this.

As Rick Archer often analogizes: "It's like we all have the winning lottery ticket sitting in our sock drawer, but we don't realize it and haven't cashed it in, so we're struggling to buy groceries and pay the rent." Eckhart Tolle echoes this sentiment in *A New Earth* (2005): "There is in our civilization a great deal of ignorance about the human condition, and the more spiritually ignorant you are, the more you suffer."[2] This ignorance is leading to self-centered behavior and rampant fear.

As I write this book, we are in the midst of a pandemic (COVID-19) that is tragically taking the world by storm. The other pandemic that is getting less attention is our collective spiritual ignorance—forgetting who and what we are; and clinging to the false, unsubstantiated belief that we are finite, separate beings.

More broadly, we've forgotten that the world we see with our eyes isn't as it seems. This "illusion" (known in Hinduism as *maya*) has persuasively led us astray. We're seeing a snake when it is in fact a rope. But when the veil thins, we start to see the world for what it is. Near the time of death, the curtain is lifted, whether we like it or not. We hear of this during well-documented "deathbed visions" and near-death experiences. Sometimes even a healthy bystander who is at the bedside of a dying person experiences it too (known as a "shared death experience").[3]

I'm often reminded of Apple cofounder Steve Jobs's famous last words. As he was dying, it seems that he "got it." He said, "Oh wow. Oh wow. Oh wow."[4] Similarly, Roger Ebert—a film critic who had doubts about whether "God" existed—saw through the veil the day before his death. "This is all an elaborate hoax" were the chilling words he wrote to his wife. He was referring to the world as an illusion. He said he "visited this other place" and saw a "vastness that you can't even imagine. It was a place where the past, present, and future were happening all at once."[5]

We're in an Evolutionary playground full of "lost souls gone wild" who are trying to find their way home. We think this apparently physical world is all there is, when in fact there is much more than meets the eye.

When all of this truly sinks in—not simply as a fun intellectual exercise but as actual reality—we are left with no choice but to reevaluate every assumption in our lives and how we live. That leads us back to where we started in the preface: setting our life's compass.

Let's begin by establishing an important framework for thinking about the basic stages of human development, as laid out by David Hawkins:[6]

○ Stage 1: What you have

- At first we focus on accumulating "things." We want a car, a house, a spouse, a job, a vacation, and so on. If we want a lot of stuff, we'll be stuck at this stage for a long time. The pandemic of spiritual ignorance keeps most of the world in this category, constantly looking for the next "thing" to try to satisfy a seemingly unfillable void. It's like a hamster running on a wheel. If we really wanted, we could skip Stage 1 after our basic needs are met. But often, the allure for "more" is too strong.

○ Stage 2: What you do

- If and when we get all the things we want, those things no longer satisfy or impress us in the same way. We move to the next stage, which is to focus on what we are perceived as *doing* in the world. We aspire to have an impressive answer to the question "What do you do?" Here, we focus on status and prestige. We say, "I'm a writer," "I'm a CEO," "I'm on the city council," and so on.

○ Stage 3: What you have become

- A focus on what we've become means that our priority becomes improving who we are in the world. As many spiritual teachers suggest, we achieve that by constantly surrendering and letting go. We stop resisting what we truly are and have always been. In essence, we employ an attitude that gets our individual ego out of the way and allows the One Mind to operate through us without obstruction.

My personal experience has taught me that "what you have" and "what you do" are dead ends, *if they're your highest priorities.* I directly experienced this in my own life and have seen it with so many others (including the rich and famous). Those phases might lead to *temporary* happiness, but it's not lasting. You'll be on to the next thing before you know it. That said, I'm not arguing that we should completely ignore "having" and "doing" and whatever fulfillment we derive from them. They might even be important aspects of our development. But if they aren't achieved within the broader context of "becoming"—which can easily happen under a physicalist belief system—we're likely to find ourselves lost in a hall of mirrors. That's where much of society is today.

And I now believe this to be true because "what you have" and "what you do" aren't in full alignment with reality. When considered in isolation, they ignore the bigger picture of the One Mind and give too much importance to egocentric desires. They place our focus on distractions and keep our eyes off the prize. Recall how Adyashanti described it during his first major awakening: **"What I had been chasing was what I am."** Our true nature as the One Mind is where love, peace, and bliss reside. That's what we're all seeking.

Furthermore, we don't take physical things and designations with us after the body dies. Rather, we bring with us the way in which we have Evolved. That is what continues beyond this physical incarnation. *Becoming* what we truly are is the name of the game, and it just so happens to be in alignment with reality. This is not what they teach us in school!

Thus, we have a chance to impact the world in a positive way as a consequence of what we are. We radiate what we *are*. For this reason, David Hawkins said: **"Our contribution to the world is the perfection of our own self."**[7] Even if "perfection" isn't possible, aiming for it is quite a life compass to set.

Why try to perfect ourselves? For one, we seem to be here to Evolve our souls. But on a more practical level, many people who achieve higher states of consciousness often demonstrate positive attributes (even if, as discussed, it's a bumpy ride getting there). Jeffery Martin has systematically studied individuals who reach a continuum of such states and has found that their lives are generally great. Those who exhibit what he calls "Fundamental Wellbeing [*sic*]" are "Finders," characterized by "high wellbeing, low neuroticism, and claim[ing] little to no depression, stress, and anxiety.... In short, the results showed that the participants were, essentially, normal and happy.... Those close to the new Finder may notice that the person has been in a great mood lately, is less reactive, and is even unusually kind and thoughtful."[8] Although Fundamental Wellbeing was arrived at through a variety of means, even including depression, "[Finders] had one thing in common: once it happened, they would not trade what they'd found for anything."[9]

Beyond personal well-being, trying to perfect ourselves has implications for the broader population. When we elevate ourselves, it's as if the entire sea level rises and all boats are lifted.[10] We contribute to the overall state of the One Mind by unifying with it, thereby *serving* it. Since the One Mind is what we essentially are anyway, by perfecting ourselves in this physical life, we also serve "our True Selves." This puts a new spin on traditional notions of "selfishness." We begin to realize that improvements we make in our own life and the lives of others ultimately help the One Mind—our highest identity. Therefore, an altruistic orientation from the Relative view is a "selfish" one from the Absolute view. Another paradox.

For this reason, author Bill McDonald suggests the approach of "waking up in the morning with a purpose, saying, 'I want to serve. Put me where I need to be.'"[11] The energy behind an intention like that is humble, powerful, and an act of love for humanity—and, by proxy, an act of love for ourselves. Could it be that life starts to serve us individually when we wholeheartedly commit to serving it?

Within this context, let's now answer the question initially posed in the preface: *What is the overall intention of your life?*

Here is the current answer for *my* life:

> **The intention of my life is to perfect myself so that I can be a pure vessel of the intelligent One Mind, thereby allowing me to serve without obstruction.**
>
> I remain open to the manner in which service can manifest—whether it seems big or small; and whether it's

through writing, personal relationships, speaking, doing business deals, cleaning dishes, or whatever.

I use my values, intuition, and passions to guide me, as they are signals that I am in alignment with the One Mind. In so doing, I focus my attention on the present moment, always doing the "next obvious thing" presented to me without over-intellectualizing projections about the future.

In order to be a pure vessel, I surrender to the "will" of the One Mind so that I align with its "current" and let it carry me rather than try to swim upstream. I maintain a "beginner's mind" and aspire to let go of my ego's attachments and arrogant judgments. This approach enables me to "wear the world like a loose garment" so that I can maintain a lighthearted and humorous attitude, enjoy life, and have fun during my time here.

Through surrender, I, as a "whirlpool," strive for liberation from the world so that I can embody the "stream" and contribute more fully *in the world* as an authentic steward of my unique skills and resources. I aspire to embody and channel the intelligence of the stream so I can optimize my service while in the whirlpool.

I aspire to be easygoing, benign, forgiving, compassionate, and unconditionally loving—toward all life, in all its expressions, without exception, including myself. My overall attitude is one of appreciation and gratitude rather than conceitedness.

I remain mindful of spiritual ego, blind spots, and glamorous temptations and seductions—requesting feedback from trusted third parties, continually trying to "clean up" and "grow up" by shedding trauma and old conditioning, and not resisting emotions that arise. I am committed to this process and aim to employ spiritual practices regularly.

Along the way, I place emphasis on taking care of my body and doing my best to be a steward of the planet (which enables us to have this Evolutionary physical experience in the first place).

An intention in this general direction can be set by anyone who's willing. You can be an athlete, a janitor, a chef, a fashion model, a teacher, a prison inmate, a politician, an Uber driver, a doctor, a parent, an investment banker—anything! What matters is the *orientation* taken to life, and the exact manifestation for each person will be what it needs to be. I'm pointing to a mind-set, an aspiration. We won't perfectly execute our intention on day one, but by passionately pointing our compass in that direction, improvement is an inevitability.

I used to be a "lost soul" just trying to stay afloat in a meaningless world. Now I feel like a chess piece in a highly meaningful cosmic scheme (even if I don't fully understand the meaning). It's as if each of us is part of a huge puzzle. The best we can do in life is to be the best version of that puzzle piece, the most authentic version of ourselves.

Pantanjali put it well: "When you are inspired by some great purpose, some extraordinary project, all your thoughts break their bonds. Your mind transcends limitations, your consciousness expands in every direction, and you find yourself in a new, great, and wonderful world. Dormant forces, faculties and talents become alive, and you discover yourself to be a greater person by far than you ever dreamed yourself to be."[12]

You never know what your impact will be, and you can't always see that impact because you don't have the Absolute perspective. For example, in life-review research, we find that *the little things are the big things*. Seemingly small acts have a ripple effect well beyond the initial action.

Let's take a look at a hypothetical situation of a poor person who feels he can't have a big impact on the world because of his finances. He adopts a compassionate life orientation and goes out of his way to strike up a pleasant conversation with a cashier at the local store. He compliments her on her job, and she acts with more positive energy throughout the day. Later in the day, because of her improved mood, she strikes up a friendlier-than-usual conversation with another customer. Unbeknownst to her, that customer runs a big corporation that is developing a cure for cancer. Because of his interaction with the cashier, he's nudged to be more productive at work that day, and as a result develops a new insight that speeds up the development of his cancer cure. The financially challenged man who thought he couldn't have an impact just did something very big without even realizing it!

This reminds me of the "butterfly effect" in Chaos Theory, which mathematically shows how tiny changes in initial conditions can lead to major, disproportionate changes in the future.[13] The analogy often used is a butterfly flapping its wings in China that can later cause a hurricane in New York (that's how the math works out). This was discovered when a meteorologist got drastically different weather predictions by adjusting one of his initial numbers by a mere decimal point. We can *be* that butterfly, effecting positive change in the world without having to know what that change will be. It all starts with how we set our life's compass.

Alternatively, maybe the fruits of our actions will be obvious. What if, by aligning with the One Mind, we start to tap into its intelligence? What if that enables creativity and *aha* moments that we can't explain, leading to new innovations? Could that allow us to bring to the world new medical treatments, technologies, and social and environmental solutions that improve life on Earth? The possibilities are endless. With radical humility, we remind ourselves that we simply don't know how powerful a shift in orientation can be.

Moreover, if the apparently "external world" is really just consciousness, then it would seem to follow that what we experience "out there" in the world is a product of what's happening in our psyche. Perhaps the state of the world today could be interpreted as the reflection of our collective internal states. If this is true, then think about how powerful it would be for a large number of people to shift their life orientations.

Recall from chapter 2 that when many people focus on a major world event, random-number generators around the world begin to act statistically nonrandomly (based on research conducted by the Global Consciousness Project led by former Princeton researcher Dr. Roger Nelson[14]). A strong collective intention in a similar direction literally shifts physical reality.

One has to wonder what the "critical mass" is: How many people would it take to shift the physical world in a noticeable way? What would be the necessary intensity level of each person's intention?

Several years ago I would have rolled my eyes at these concepts, thinking they sounded like idealistic, wishful thinking. But look at what's happening in the world today: our structures are simply falling apart everywhere we turn, and our environment is collapsing. Most of the world is operating from a compass that is completely misaligned with reality.

We can't hide our heads in the sand anymore if we want our civilization to exist on this planet. **Something big needs to change.** To me, drastically shifting our collective orientation is the only solution. Nothing else hits the root issue. If we don't align with reality, reality will continue to rock us. The jig is up. Time is running out, whether we want to ignore it or not.

The choice resides within each of us during this inflection point in human history. **You have the power to set your life's compass, no matter who you are.** So, now I ask you again what I asked at the start of this book: *What is the overall intention of your life?* And, most important: *What are you going to do about it?*

If we do shift our compasses, how will we then run nations? How will we run businesses? How will we manage our day-to-day relationships? How will we treat animals? How will we treat the environment? How will we build a "New Earth" based on spiritual development and loving service?

It's time for us to wake up, clean up, and grow up. The survival of the species depends on it.

ACKNOWLEDGMENTS

Thank you to my literary agent/publisher Bill Gladstone of Waterside Productions, and his wonderful wife, Gayle, for being so supportive and publishing this book so quickly. Waterside has played an enormous role in helping me get my messages out to the world.

I feel fortunate to have worked again with such a fantastic editor in Kenneth Kales and talented proofreader in Jill Kramer. Thank you to Ken Fraser for another excellent cover design and to Joel Chamberlain for another great job with typesetting and image design. And thanks to Jennifer Uram for assistance with my contracts.

This book was only possible because of the heroic work of those who came before me, many of whom are referenced in this book. Without the work of these brilliant and brave scientists, philosophers, and spiritual teachers, I would not be where I am today. It is because of *them* that I'm able to write books like this, thereby serving others.

Additionally, as I mentioned in this book, the *Buddha at the Gas Pump* podcast has had a major influence on me as well. I regard it as one of the greatest gifts to humanity that I've come across. On behalf of the world, I thank Rick and Irene Archer and the full *Buddha at the Gas Pump* team. My consciousness has been elevated significantly from what I've absorbed by listening to that show, and I know the same is (and will be) true for countless others.

A huge thanks to Dr. Eben Alexander, Dr. Ervin Laszlo, Dr. Larry Dossey, Dr. Cassi Vieten, Rick Archer, and Rev. Bill McDonald for their endorsements on such short notice. All of them have had a significant influence on my thinking, so I'm honored that they were willing to provide support for this book.

I also thank Rick Archer for providing incredibly thoughtful and helpful feedback as he read my manuscript. My friends Nancy Heydemann, Natalie Carlstead, Callie Kron, and Matt Ford are saints for agreeing to read the book at the last minute and provided great feedback as well.

In my previous book, I tried to list as many friends as I could think of who supported me through my journey over the past several years. Inevitably, I later felt guilty about having accidentally omitted names, so to avoid that, here I will say to all of my friends (you know who you are): *Thank you!* It's been so terrific having support from all angles. And, of course, I thank everyone who has taken a role in my "army of counselors" (you know who you are as well) for keeping me in check.

Thank you to my fantastic, supportive former colleagues at Sherpa Technology Group, particularly Kevin Rivette, Ralph Eckardt, Peter Detkin, Andy Filler, Calvin Wong, Michael Poppler, Colin Santangelo, Matt Mahoney, Kevin Mills, AJ Joshi, and Nicole Neville.

My podcast producers and friends, Matt Ford and Gabe Goodwin, have been amazing during this journey. Thank you for everything.

I also thank my parents, Karen and Bob; and my two younger brothers, Zack and Jake, for their unconditional support and love.

Finally, I thank my appreciative supporters who have given me terrific feedback on my first book, podcast, and interviews. It always means a lot to me!

ENDNOTES

Chapter 1

1   Dawkins, *The Selfish Gene*, 15.

2   As cited in Popova, *The Unity of the Universe: Nobel-Winning Physicist Steven Weinberg on Simplicity and Complexity, Science and Religion, and the Mother of All Questions*. https://www.brainpickings.org/2017/02/01/we-are-all-stardust-steven-weinberg-interview/.

3   Carroll, *The Big Picture*, 431.

4   Tsakiris, *This prominent scientist says life is meaningless... and he's serious*, https://skeptiko.com/sean-carroll-thinks-life-is-meaningless-314/.

5   As cited in Popova, *The Unity of the Universe: Nobel-Winning Physicist Steven Weinberg on Simplicity and Complexity, Science and Religion, and the Mother of All Questions*. https://www.brainpickings.org/2017/02/01/we-are-all-stardust-steven-weinberg-interview/.

6   This triangle and others in the book have been adapted from the work of Dr. Dean Radin (for example, see his book *Real Magic*).

Chapter 2

1   Kastrup, *Why Materialism Is Baloney*, 31.

2   Harris, *Waking Up*, 60.

3   *Science* magazine's 125th-anniversary issue, http://www.sciencemag.org/site/feature/misc/webfeat/125th/. The question is phrased as follows: "What is the Biological Basis of Consciousness?"

4   Kastrup, *Transcending the Brain*, https://blogs.scientificamerican.com/guest-blog/transcending-the-brain/.

5   For more, see *Ep 2. Blindfolded* of *Where Is My Mind?* podcast and chapter 2 of *An End to Upside Down Thinking*.

6   Treffert, *Islands of Genius*, 122.

7   See *Ep 2. Blindfolded* of *Where Is My Mind?* podcast.

8   See chapter 2 of *An End to Upside Down Thinking*.

9   Ibid.

10   Reville, *Remarkable story of maths genius who had almost no brain*.

11   Pietsch, *Shufflebrain*, 1.

12   See Kastrup, *Transcending the Brain*, https://blogs.scientificamerican.com/guest-blog/transcending-the-brain/.

13   Ibid.

14   Ibid.

15   Ibid.

16   Interview with Max Planck, *The Observer*.

17   Schrödinger, *What Is Life? with Mind and Matter*, 139.

18   See Kastrup, *Why Materialism Is Baloney*.

19   See Kastrup, *The Idea of the World*.

20   "Lawrence Krauss – Does ESP Make Sense?", https://www.youtube.com/watch?v=5NweHLQmbZE.

21   See chapter 4 of *An End to Upside Down Thinking* and *Ep 4 CIA Psychic Spying and Knowing the Future* of *Where Is My Mind?* podcast.

22   See chapter 5 of *An End to Upside Down Thinking* and *Ep 3 Telepathy* of *Where Is My Mind?* podcast.

23   See chapter 6 of *An End to Upside Down Thinking* and *Ep 4 CIA Psychic Spying and Knowing the Future* of *Where Is My Mind?* podcast.

24   See chapter 8 of *An End to Upside Down Thinking* and *Ep 8 Revolution* of *Where Is My Mind?* podcast. Also see *Connected* by Dr. Roger Nelson.

25   For a discussion on "six sigma" statistical results, see Radin, *Real Magic*, 97. Also see Radin, *Supernormal, Entangled Minds*, and *The Conscious Universe*. Additional resources include: Tressoldi, *Extraordinary claims require extraordinary evidence: The case of non-local perception, a classical and Bayesian review of evidences*; Williams, *Revisiting the ganzfeld ESP debate: A basic review and assessment*; Mossbridge, Tressoldi, & Utts, *Predictive physiological anticipation preceding seemingly unpredictable stimuli: A meta-analysis*; Bem, Tressoldi, Raberyon, & Duggan, *Feeling the future: A meta-analysis of 90 experiments on the anomalous anticipation of random future events*; Bosch, Steinkamp, & Boller, *Examining psychokinesis: The interaction of human intention with random number generators—a meta-analysis*; Radin, Nelson, Dobyns, & Houtkooper, *Re-examining psychokinesis: Commentary on the Bösch, Steinkamp and Boller meta-analysis*; Nelson, Radin, Shoup, & Bancel, *Correlations of continuous random data with major world events*. Also see global-mind.org/results.html.

26   Utts, *An Assessment of the Evidence for Psychic Functioning*.

27   Cardeña, *The Experimental Evidence for Parapsychological Phenomena:*

A Review.

28   Carey, *A Princeton Lab on ESP Plans to Close its Doors*, http://www. nytimes. com/2007/02/10/science/10princeton.html?mcubz=0.

29   See chapter 7 of *An End to Upside Down Thinking* and *Ep 3 Telepathy* of *Where Is My Mind?* podcast.

30   See *Ep 5 Near-Death Experiences* of *Where Is My Mind?* and *The Self Does Not Die*, 58–59.

31   For more on near-death experiences, see chapter 9 of *An End to Upside Down Thinking* and *Ep 5 Near-Death Experiences* of *Where Is My Mind?* podcast.

32   For more on mediumship and after-death communications, see chapter 10 of *An End to Upside Down Thinking* and *Ep 7 Mediumship and Reincarnation* of *Where Is My Mind?* podcast.

33   For more on reincarnation research, see chapter 11 of *An End to Upside Down Thinking* and *Ep 7 Mediumship and Reincarnation* of *Where Is My Mind?* podcast.

34   Cardeña, *Eminent People Interested in Psi.*

35   Strassman, *DMT: The Spirit Molecule*, 234.

36   Greenwell, *When Spirit Leaps*, vii.

37   Van Lommel, *Near-Death Experience, Consciousness, and the Brain.*

38   See chapter 3 of *An End to Upside Down Thinking* for further details on physics.

39   As cited in Russell, *From Science to God*, 49.

40   Heisenberg, *Physics and Philosophy: The Revolution in Modern Science*, 80–81.

41   Kastrup, *Brief Peeks Beyond*, 36.

42   Kastrup, *Why Materialism Is Baloney*, 62.

43   This argument is a summary of the case made by Dr. Bernardo Kastrup in chapter 3 of *Why Materialism Is Baloney.*

44   Gosling, *Science and the Indian Tradition: When Einstein Met Tagore*, 162.

45   Baruss and Mossbridge, *Transcendent Mind*, 25.

46   Gober, *An End to Upside Down Thinking*, 240. I use the term "finite and separate," which comes from Rupert Spira's teaching, as does the "symptoms/disease" metaphor.

Chapter 3

1   Carroll, *The Big Picture*, 430.

2   Dawkins, *The God Delusion*, 414.

3   *Cosmos in You* podcast interview with Donald Hoffman, July 27, 2015, https://www.suzannahscully.com/full-episodes.

4   A term often used by Dr. David R. Hawkins, e.g., *The Eye of the I*, 148.

5   For more on this topic, see Rick Archer's talk at the 2019 Science and Nonduality conference entitled "Knowledge/Reality is Different in Different States of Consciousness," https://batgap.com/

rick-archer-sand-conference-knowledge-reality-different-states-consciousness/.

6   Kastrup, *The threat of panpsychism: a warning*, https://www.bernardokastrup.com/2015/05/the-threat-of-panpsychism-warning.html.

7   Goff, Philip. Twitter post. January 27, 2020. https://twitter.com/Philip_Goff/status/1221778157935697920.

8   Spira, *The Nature of Consciousness*, 31.

9   Discussed further in chapter 13 of *An End to Upside Down Thinking*. The term *modulations of consciousness* is often used by Rupert Spira.

10   Segal, *Collision with the Infinite*, 161.

11   Dyer, *Getting in the Gap*, 31.

12   Lanza and Berman, *Biocentrism*, 84–85.

13   As stated in *Biocentrism*, 85-86: "Values given are from the CODATA 1998 recommended by the National Institute of Standards and Technology of the United States (NIST). Values contain the (uncertainty) in the last two decimals places given in brackets. Values that do not have this uncertainty listed are exact. For example: m_u=1.66053873(13) x 10^{-27} kg; m_u=1.66053873 x 10^{-27} kg; Uncertainty in m_u=0.00000013 x 10^{-27} kg."

14   Ibid., 91–92.

15   Kastrup, *Brief Peeks Beyond*, 113.

16   Ibid., 113–119.

17   Hawkins, *Letting Go*, 191–194.

18   *Where Is My Mind?* podcast *Ep 6 The Life Review*.

19   Ring, *The Golden Rule Drastically Illustrated*, 11–12 from Sunfellow, *The Purpose of Life*.

20   *Where Is My Mind?* podcast *Ep 6 The Life Review*.

21   Ibid.

22   *The Spiritual Teaching of Ramana Maharshi*, 59.

23   The term *Evolutionary impulse* was often used by the late Barbara Marx Hubbard.

24   Caplan, *Eyes Wide Open*, xxx.

25   See *Where Is My Mind?* podcast, *Ep 5 Near-Death Experiences* and *Ep 8 Revolution*.

26   Sagan, *The Demon-Haunted World*, 302.

27   Shermer, *Heavens on Earth*, 98–99.

28   Sharma and Tucker, *Cases of the Reincarnation Type with Memories from Intermission Between Lives*.

29   Matlock, *Life planning during the intermission*.

30   Atwater, *The Forever Angels*, 13.

31   Stevenson, *Children Who Remember Previous Lives*, 46.

32   Schwartz, *Your Soul's Plan*, 31.

33   *Buddha at the Gas Pump* podcast episode with Noah Elkrief, September 16, 2019, https://batgap.com/noah-elkrief/.

34   *DMT: The Spirit Molecule* documentary.

35   Hawkins, *Letting Go*, 193.

36   Lorimer, *Prophet of Our Times*, 50.

37   *Buddha at the Gas Pump* podcast episode with Andrew Harvey, January
       31, 2016, https://batgap.com/andrew-harvey/.

Chapter 4

1   Adapted from Suzanne Segal's advice to "follow the obvious" in *Collision
       with the Infinite*, 163.
2   Hawkins, *I: Reality and Subjectivity*, 423.
3   Hawkins, *Letting Go*, 19–20.
4   *What Is Affective Foreceasting?*, https://www.psychologytoday.com/us/
       basics/affective-forecasting.
5   *Buddha at the Gas Pump* podcast episode with Isira, February 10, 2020,
       https://batgap.com/isira/.
6   *Buddha at the Gas Pump* podcast episode with Julie Chimes, June 10,
       2011, https://batgap.com/julie-chimes/.
7   Ritvo, *The Neuroscience of Giving*, https://www.psychologytoday.com/us/
       blog/vitality/201404/the-neuroscience-giving.
8   Hawkins (Ed. Scott Jeffrey), *Along the Path to Enlightenment*, 181.
9   *Buddha at the Gas Pump* podcast episode with Pernilla Lillarose, May 21,
       2017, https://batgap.com/pernilla-lillarose/.
10   Anita Moorjani Facebook post, January 6, 2020. https://www.facebook.
       com/Anita.Moorjani/photos/a.252465361465136/3116167578428219/
       ?type=3&eid=ARBiGFK0ouSQxrCuPsPB0gw9zyrntLdRScGJ1MDI-
       Djx2pIdinDlM7pWYmn10wsqYNWooBnOGzafYppZq.
11   I've heard this analogy used on a number of occasions, including in
       teachings by Dr. David Hawkins.
12   Saraswati, *Value of Values*, 33.
13   Corazza, *An Introduction to Maharishi Vedic Science*.
14   Kaplan, *Jewish Meditation*, 165.

Chapter 5

1   Greenwell, *When Spirit Leaps*, 44–82.
2   Sri Sri Ravi Shankar, *Meditation and Karma*, https://www.artofliving.
       org/us-en/meditation/srisri-meditation/meditation-karma.
3   Greenwell, *When Spirit Leaps*, 9.
4   Ibid., 9–10.
5   *conscioustv* interview with Adyashanti, August 5, 2011, https://www.you-
       tube.com/watch?v=lUTF8n_WJko.
6   Hawkins, *I: Reality and Subjectivity*, 378–380.
7   Greenwell, *When Spirit Leaps*, 10.
8   Personal correspondence with Rick Archer, April 2, 2020.
9   Greenwell, *The Kundalini Guide*, 19.
10   In this section, I refer to the term "kundalini energy," which describes an
       often-discussed phenomenon among spiritual seekers. I wonder, however,
       if that general category of energy has multiple flavors, or alternatively

if there are different types of "energies" that we can experience. In my own life, some of the energies I've felt have resulted in different sensations. Were all of those experiences versions of kundalini energy, or were they different energies altogether? Since the experience of these energies is inherently subjective, and difficult to describe, it can be challenging to categorize them. I mention this because the discussion of palpable "energy" might be incredibly nuanced, and at this point I simply don't know enough to accurately categorize the potentially different types. Therefore, this section focuses on the most commonly discussed energy category I've encountered (kundalini), which I suspect has at least some relation to other energies.

11   Greenwell, *The Kundalini Guide*, 39–64.
12   Hawkins, *I: Reality and Subjectivity*, xxvi.

Chapter 6

1   As cited in Caplan, *Eyes Wide Open*, 146–147.
2   As cited in Ibid., 147.
3   As cited in Ibid., 148.
4   Ibid., 19–20.
5   Ibid., 36.
6   Ibid., 110.
7   *Buddha at the Gas Pump* episode with Ken Wilber, April 30, 2018, https://batgap.com/ken-wilber/.
8   Ibid.
9   Henriques, *Can the legacy of trauma be passed down the generations?*, https://www.bbc.com/future/article/20190326-what-is-epigenetics.
10   Caplan, *Eyes Wide Open*, 139.
11   Ibid., 10–11.
12   Ibid., 43.
13   Radin, *Supernormal*, 61–63 and *Macro-Psychokinesis, Part One: Physical Mediumship* with Stephen E. Braude on *New Thinking Allowed* with Jeff Mishlove.
14   Yogananda, *Autobiography of Yogi*, 115.
15   My full-length interview with Wahbeh is available to subscribers at www.markgober.com/podcast along with dozens of other interviews I conducted.
16   *Buddha at the Gas Pump* episode with Gina Lake, March 29, 2012, https://batgap.com/gina-lake/.
17   See Swami Vivekananda's commentary (available at the following link, for example: https://universaltheosophy.com/pdf-library/Yoga%20Aphorisms_SV.pdf): "The Yogi should not feel allured or flattered by the overtures of celestial beings for fear of evil again. There are other dangers too; gods and other beings come to tempt the Yogi. They do not want anyone to be perfectly free. They are jealous, just as we are, and worse than us

sometimes. They are very much afraid of losing their places. Those Yogis who do not reach perfection die and become gods; leaving the direct road they go into one of the side streets, and get these powers. Then, again, they have to be born. But he who is strong enough to withstand these temptations and go straight to the goal, becomes free."

18 Caplan, *Eyes Wide Open*, 247.

19 McDaniels, *Psychedelics reduce anxiety, depression in patients, study finds*, http://www.baltimoresun.com/health/bs-hs-psychedelics-cancer-2016 1201-story.html.

20 Greenwell, *When Spirit Leaps*, 63.

21 Backster, *Primary Perception*, 24–25.

Chapter 7

1 Laszlo, *The Intelligence of the Cosmos*, 45.

2 Tolle, *A New Earth*, 284.

3 See chapters 9 and 10 in *An End to Upside Down Thinking*.

4 Jones, *Steve Jobs's last words: 'Oh wow. Oh wow. Oh wow' Jobs*, 2011. https://www.theguardian.com/technology/2011/oct/31/steve-jobs-last-words.

5 Ebert, *Reflecting for the New Year: On Roger's Last Day*, https://www.rogerebert.com/chazs-blog/reflecting-for-the-new-year-on-rogers-last-day.

6 "Dr. David Hawkins talk at Unity Marin", September 25, 2004, https://www.youtube.com/watch?v=tPz96mV93io.

7 Hawkins, "Transcending the Levels of Consciousness: A review of the work, 09/06," Audible chapter 8.

8 Martin, *The Finders*, 30–31.

9 Ibid., 20.

10 This was a metaphor often used by David Hawkins.

11 *The near-death experiences of Bill McDonald*, Nov 6, 2019, https://www.youtube.com/watch?v=4xrtyOqwr9E.

12 Dyer, *Where Do You Live?*, https://www.drwaynedyer.com/blog/tag/inspired/.

13 See chapter 3 of *An End to Upside Down Thinking*.

14 For more on this topic, see *Connected* by Dr. Roger Nelson and *Ep 8 Revolution* of *Where Is My Mind?* podcast.

BIBLIOGRAPHY

"Affective Forecasting." *Psychology Today* website, n.d. https://www.psychologytoday.com/us/basics/affective-forecasting.

Anthony Chene production. "The near-death experiences of Bill McDonald." YouTube video, November 6, 2019, https://www.youtube.com/watch?v=4xrtyOqwr9E.

Archer, Rick. "517. Noah Elkrief." Buddha at the Gas Pump website, September 16, 2019.

Archer, Rick. "530. Rick Archer at the 2019 SAND Conference—Knowledge/Reality is Different in Different States of Consciousness." Buddha at the Gas Pump website, January 1, 2020.

Archer, Rick. "536. Isira." Buddha at the Gas Pump website, February 10, 2020.

Archer, Rick. "452. Ken Wilber." Buddha at the Gas Pump website, April 30, 2018.

Archer, Rick. "401. Pernilla Lillarose." Buddha at the Gas Pump website, May 21, 2017.

Archer, Rick. "116. Gina Lake." Buddha at the Gas Pump website, March 29, 2012.

Archer, Rick. "071. Julie Chimes." Buddha at the Gas Pump website, June 10, 2011.

Archer, Rick. "329. Andrew Harvey." Buddha at the Gas Pump website, January 31, 2016.

Atwater, P.M.H. *The Forever Angels: Near-Death Experiences in Childhood and Their Lifelong Impact*. Rochester: Bear & Company, 2019.

Backster, Cleve. *Primary Perception: Biocommunication with Plants, Living Foods, and Human Cells*. Anza, CA: White Rose Millennium, 2003.

Bem, Daryl, et al. "Feeling the Future: A Meta-Analysis of 90 Experiments on the Anomalous Anticipation of Random Future Events." *F1000Research* 4 (2015): 1188. doi: 10.12688/f1000research.7177.1) https://f1000research. com/articles/4-1188/v1.

Bigelow, Ben. "Dr. David Hawkins talk at Unity Marin." YouTube Video, September 25, 2004, https://www.youtube.com/watch?v=tPz96mV93io.

Bosch, Holger, Fiona Steinkamp, and Emil Boller. "Examining Psychokinesis: The Interaction of Human Intention with Random Number Generators; A Meta-Analysis." *Psychological Bulletin* 132, no. 4 (2006): 497–523.

Baruŝs, Imants, and Julia Mossbridge. *Transcendent Mind: Rethinking the Science of Consciousness*. Washington, DC: American Psychological Association, 2017.

Caplan, Mariana. *Eyes Wide Open: Cultivating Discernment on the Spiritual Path*. Boulder: Sounds True, 2009.

Cardeña, Etzel. "Eminent People Interested in Psi." *Psi Encyclopedia*. London: The Society for Psychical Research. January 10, 2020 (last updated).

Cardeña, Etzel. (2018). "The experimental evidence for parapsychological phenomena: A review." *American Psychologist, 73*(5), 663–677. https://doi.org/10.1037/amp0000236.

Carey, Benedict. "A Princeton Lab on ESP Plans to Close Its Doors." *New York Times*, February 10, 2007. http://www.nytimes.com/2007/02/10/science/10princeton.html?mcubz=0.

Carroll, Sean. *The Big Picture: On the Origins of Life, Meaning, and the Universe Itself.* New York: Dutton, 2016.

Closer to Truth. "Lawrence Krauss: Does ESP Make Sense?" YouTube Video, July 17, 2017. https://youtu.be/5NweHLQmbZE.

conscioustv. Adyashanti - 'Awakening' - interview by Renate McNay. YouTube Video, August 5, 2011. https://www.youtube.com/watch?v=lUTF8n_WJko.

Corazza, Paul. "An Introduction to Maharishi Vedic Science." 1993. http://pcorazza.lisco.com/papers/MVS/IntroToMVS.pdf.

Dawkins, Richard. *The God Delusion.* New York: First Mariner, 2008.

Dawkins, Richard. *The Selfish Gene.* New York: Oxford University Press, 1989.

Dyer, Wayne. *Getting in the Gap.* Carlsbad, CA: Hay House, 2003.

Dyer, Wayne. "Where Do You Live?" *Dr. Wayne Dyer* website, n.d., https://www.drwaynedyer.com/blog/tag/inspired/.

Ebert, Chaz. "Reflecting for the New Year: On Roger's Last Day." *Chaz's Journal* website, January 1, 2014. https://www.rogerebert.com/chazs-blog/reflecting-for-the-new-year-on-rogers-last-day.

"Formal Results: Testing the GCP Hypothesis." Global Consciousness Project website, n.d. global-mind.org/results.html.

Ginko Balboa. "DMT: The Spirit Molecule (2010) [multi subs]." YouTube Video, Nov 23, 2016, https://www.youtube.com/watch?v=fwZqVqbkyLM.

Gober, Mark. *An End to Upside Down Thinking: Dispelling the Myth That the Brain Produces Consciousness, and the Implications for Everyday Life.* Cardiff-by-the-Sea, CA: Waterside Press, 2018.

Gober, Mark and Blue Duck Media. *Where Is My Mind?* podcast (Episodes 1 through 8). June–September 2019.

Goff, Philip. Twitter post. January 27, 2020. https://twitter.com/Philip_Goff/status/1221778157935697920.

Gosling, David. *Science and the Indian Tradition: When Einstein Met Tagore.* London: Routledge, 2007.

Greenwell, Bonnie. *The Kundalini Guide: A Companion for the Inward Journey.* Ashland, OR: Shakti River, 2014.

Greenwell, Bonnie. *When Spirit Leaps: Navigating the Process of Spiritual Awakening*. Oakland, CA: Non-Duality Press, 2018.

Harris, Sam. *Waking Up: A Guide to Spirituality Without Religion*. New York: Simon & Schuster, 2015.

Hawkins, David. "December 6." In *Along the Path to Enlightenment*, edited by Scott Jeffrey. Carlsbad, CA: Hay House, 2011.

Hawkins, David. *I: Reality and Subjectivity*. Alexandria, Australia: Hay House, 2003.

Hawkins, David. *Letting Go: The Pathway of Surrender*. Alexandria, Australia: Hay House, 2014.

Hawkins, David. *Transcending the Levels of Consciousness: A Review of the Work, 09/06*. Audible Chapter 8. Veritas: 2006.

Hawkins, David. *The Eye of the I: From Which Nothing Is Hidden*. Carlsbad, CA: Hay House, 2001.

Heisenberg, Werner. *Physics and Philosophy: The Revolution in Modern Science*. New York: HarperPerennial, 1958.

Henriques, Martha. "Can the legacy of trauma be passed down the generations?" *BBC Future*, March 26, 2019. https://www.bbc.com/future/article/20190326-what-is-epigenetics.

"Interview with Max Planck." *The Observer*, January 25, 1931.

Jones, Sam. "Steve Jobs's last words: 'Oh wow. Oh wow. Oh wow.'" *The Guardian* website, Oct 31, 2011. https://www.theguardian.com/technology/2011/oct/31/steve-jobs-last-words.

Kaplan, Aryeh. *Jewish Meditation: A Practical Guide*. New York: Schocken Book, 1985.

Kastrup, Bernardo. *Brief Peeks Beyond: Critical Essays on Metaphysics, Neuroscience, Free Will, Skepticism, and Culture*. Winchester, UK: Iff, 2015.

Kastrup, Bernardo. "Transcending the Brain: At Least Some Cases of Physical Damage Are Associated with Enriched Consciousness or Cognitive Skill." *Scientific American* blog, March 29, 2017. https://blogs.scientificamerican.com/guest-blog/transcending-the-brain/.

Kastrup, Bernardo. "The threat of panpsychism: a warning." *Metaphysical Speculations* blog, May 24, 2015. https://www.bernardokastrup.com/2015/05/the-threat-of-panpsychism-warning.html.

Kastrup, Bernardo. *Why Materialism Is Baloney: How True Skeptics Know There Is No Death and Fathom Answers to Life, the Universe and Everything.* Winchester, UK: Iff, 2014.

Kastrup, Bernardo. *The Idea of the World: A Multi-Disciplinary Argument for the Mental Nature of Reality.* Hampshire, UK: Iff, 2019.

Lanza, Robert, and Bob Berman. *Biocentrism: How Life and Consciousness Are the Keys to Understanding the True Nature of the Universe.* Dallas, TX: BenBella, 2009.

Laszlo, Ervin. *The Intelligence of the Cosmos: Why Are We Here? New Answers from the Frontiers of Science.* Rochester, VT: Inner Traditions, 2017.

Lorimer, David. "God." In *Prophet for Our Times: The Life & Teachings of Peter Deunov"* edited by David Lorimer. Carlsbad, CA: Hay House, 2015.

Maharshi, Ramana. *The Spiritual Teaching of Ramana Maharshi.* Boulder: Shambhala Classics, 2004.

Martin, Jeffery. *The Finders.* Jackson: Integration Press, 2019.

Matlock, James. "Life planning during the intermission." *Signs of Reincarnation*, April 18, 2017. http://jamesgmatlock.com/2017/04/18/life-planning-during-the-intermission/.

McDaniels, Andrea. "Psychedelics Reduce Anxiety, Depression in Patients, Study Finds." Baltimore Sun, December 1, 2016. http://www.baltimoresun.com/ health/bs-hs-psychedelics-cancer-20161201-story.html.

Moorjani, Anita. Facebook post. January 6, 2020. https://www.facebook.com/Anita.Moorjani/photos/a.252465361465136/3116167578428219/?-type=3&eid=ARBiGFK0ouSQxrCuPsPB0gw9zyrntLdRScGJ1MDIDjx2pIdinDlM7pWYmn10wsqYNWooBnOGzafYppZq.

Mossbridge J., P. Tressoldi, and J. Utts. "Predictive Physiological Anticipation Preceding Seemingly Unpredictable Stimuli: A Meta-Analysis." *Frontiers in Psychology* 3 (2012): 390. https://www.frontiersin.org/articles/10.3389/ fpsyg.2012.00390/full.

Nelson, R.D. et al. "Correlations of Continuous Random Data with Major World Events." *Foundations of Physics Letters* 15 (2002): 537-50.

Nelson, Roger. *Connected*. Princeton: ICRL Press, 2019.

New Thinking Allowed with Jeffrey Mishlove. "Macro-Psychokinesis, Part One: Physical Mediumship, with Stephen E. Braude," YouTube Video, September 9, 2015. https://www.youtube.com/watch?v=2GtXY3xhmag.

"125th Anniversary Issue." *Science* magazine website. http://www.sciencemag. org/site/feature/misc/webfeat/125th/.

Pearson, Craig. *The Supreme Awakening: Experiences of Enlightenment Throughout Time—and How You Can Cultivate Them*. Fairfield: Maharishi University of Management Press, 2013.

Pietsch, Paul. *Shufflebrain*. Boston: Houghton Mifflin, 1981.

Popova, Mary. "The Unity of the Universe: Nobel-Winning Physicist Steven Weinberg on Simplicity and Complexity, Science and Religion, and the Mother of All Questions." *Brain Pickings*, n.d. https://www.brainpickings.org/2017/02/01/we-are-all-stardust-steven-weinberg-interview/.

Radin, Dean. *The Conscious Universe: The Scientific Truth of Psychic Phenomena*. New York: HarperEdge, 1997.

Radin, Dean. *Entangled Minds Extrasensory Experiences in a Quantum Reality*. New York: Paraview, 2006.

Radin, Dean, et al. "Re-Examining Psychokinesis: Commentary on the Bösch, Steinkamp, and Boller Meta-Analysis." *Psychological Bulletin* 132 (2006): 529–32.

Radin, Dean. *Real Magic: Ancient Wisdom, Modern Science, and a Guide to the Secret Power of the Universe*. New York: Harmony, 2018.

Radin, Dean. *Supernormal: Science, Yoga, and the Evidence for Extraordinary Psychic Abilities*. New York: Random House, 2013.

Reville, William. "Remarkable story of maths genius who had almost no brain." *Irish Times*, November 9, 2006. https://www.irishtimes.com/news/remarkable-story-of-maths-genius-who-had-almost-no-brain-1.1026845.

Ring, Kenneth. "The Golden Rule Dramatically Illustrated." In *The Purpose of Life: As revealed by near-death experiences from around the world*, by David Sunfellow, 2019.

Ritvo, Eva. "The Neuroscience of Giving: Proof that helping others helps you." *Psychology Today*, April 24, 2014. https://www.psychologytoday.com/us/blog/vitality/201404/the-neuroscience-giving.

Rivas, Titas, Anny Dirven, and Rudolf H. Smit. *The Self Does Not Die: Verified Paranormal Phenomena from Near-Death Experiences.* Durham: IANDS Publications, 2016.

Russell, Peter. *From Science to God: The Mystery of Consciousness.* Novato, CA: New World Library, 2003.

Sagan, Carl. *The Demon-Haunted World: Science As a Candle in the Dark.* New York: Random House, 1995.

Saraswati, Swami Dayananda. *Value of Values.* Mylapore, India: Arsha Vidya Research and Publication Trust, 2007.

Schrödinger, Erwin. *What Is Life? With Mind and Matter and Autobiographical Sketches.* London: Cambridge University Press, 1969.

Schwartz, Robert. *Your Soul's Plan: Discovering the Real Meaning of the Life You Planned Before You Were Born.* Berkeley: Frog Books, 2009.

Scully, Suzannah. "Donald Hoffman: Do We See Reality As It Really Is?." Cosmos in You website, July 27, 2015. https://www.suzannahscully.com/ full-episodes.

Segal, Suzanne. *Collision with the Infinite: A Life Beyond the Personal Self.* San Diego, CA: Blue Dove Press, 1996.

Shankar, Sri Sri Ravi. *Meditation and Karma.* Art of Living website, n.d., https://www.artofliving.org/us-en/meditation/srisri-meditation/meditation-karma.

Sharma, Poonam and Jim Tucker, MD. "Cases of the Reincarnation Type with Memories from the Intermission Between Lives." *Journal of Near-Death Studies* 23 (2) (2004), 101–118.

Shermer, Michael. *Heavens on Earth: The Scientific Search for the Afterlife, Immortality, and Utopia.* New York: Henry Holt and Company, 2018.

Spira, Rupert. *The Nature of Consciousness: Essays on the Unity of Mind and Matter*. Oakland, CA: New Harbinger, 2017.

Stevenson, Ian. *Children Who Remember Previous Lives: A Question of Reincarnation*. Jefferson, NC: McFarland, 2001.

Strassman, Rick. *DMT: The Spirit Molecule: A Doctor's Revolutionary Research into the Biology of Near-Death and Mystical Experiences*. Rochester: Park Street Press, 2001.

Treffert, Darold. *Islands of Genius: The Bountiful Mind of the Autistic, Acquired, and Sudden Savant*. London: Jessica Kingsley, 2012.

Tressoldi, P. E. "Extraordinary Claims Require Extraordinary Evidence: The Case of Non-Local Perception, a Classical and Bayesian Review of Evidences." *Frontiers in Psychology* 2, no. 117 (2011).

Tsakiris, Alex. "This prominent scientist says life is meaningless . . . and he's serious." *Skeptiko* website, May 10, 2016. https://skeptiko. com/sean-carroll-thinks-life-is-meaningless-314/.

Tolle, Eckhart. *A New Earth: Awakening to Your Life's Purpose*. New York: Penguin Books, 2005.

Utts, Jessica. "An Assessment of the Evidence for Psychic Functioning." *Journal of Parapsychology* 59, no. 4 (1995): 289–320.

van Lommel, Pim. "Near-Death Experience, Consciousness, and the Brain: A New Concept about the Continuity of Our Consciousness Based on Recent Scientific Research on Near-Death Experience in Survivors of Cardiac Arrest." *World Futures* 62 (2006): 134–51. http://deanradin.com/evidence/vanLommel2006.pdf.

Wahbeh, H., Carpenter, L., Radin, D. "A mixed methods phenomenological and exploratory study of channeling." *Journal of the Society for Psychical Research* 82, no. 3 (2018): 129–148.

Williams, B. J. "Revisiting the Ganzfeld ESP Debate: A Basic Review and Assessment." *Journal of Scientific Exploration* 25, no. 4 (2011): 639–61.

Yogananda, Paramahansa. *Autobiography of a Yogi*. UK: Rider, 1955.

INDEX

ABOUT THE AUTHOR

 Mark Gober is an international speaker, author of the award-winning book *An End to Upside Down Thinking* (2018), and host of the podcast *Where Is My Mind?* (2019). Additionally, he serves on the Board of the Institute of Noetic Sciences and the School of Wholeness and Enlightenment. Mark's background is in business as a Partner at Sherpa Technology Group in Silicon Valley and previously as an investment banking analyst in New York. He graduated *magna cum laude* from Princeton University, where he wrote his award-winning thesis on Daniel Kahneman's Nobel Prize–winning "Prospect Theory" and was elected a captain of Princeton's Division I Tennis Team.

www.markgober.com